Soil, Water and Nitrates Management in Horticultural Production

Soil, Water and Nitrates Management in Horticultural Production

Editor

Rui Manuel Almeida Machado

MDPI • Basel • Beijing • Wuhan • Barcelona • Belgrade • Manchester • Tokyo • Cluj • Tianjin

Editor
Rui Manuel Almeida Machado
Universidade de Évora
Portugal

Editorial Office
MDPI
St. Alban-Anlage 66
4052 Basel, Switzerland

This is a reprint of articles from the Special Issue published online in the open access journal *Horticulturae* (ISSN 2311-7524) (available at: https://www.mdpi.com/journal/horticulturae/special_issues/Management_Horticultural_Production).

For citation purposes, cite each article independently as indicated on the article page online and as indicated below:

LastName, A.A.; LastName, B.B.; LastName, C.C. Article Title. *Journal Name* **Year**, *Volume Number*, Page Range.

ISBN 978-3-0365-4575-2 (Hbk)
ISBN 978-3-0365-4576-9 (PDF)

Contents

Editorial

Soil, Water and Nitrates Management in Horticultural Production

Rui Manuel Almeida Machado

MED—Mediterranean Institute for Agriculture, Environment and Development, Departamento de Fitotecnia, Escola de Ciências e Tecnologia, Universidade de Évora, Pólo da Mitra, Ap. 94, 7006-554 Évora, Portugal; rmam@uevora.pt

Citation: Manuel Almeida Machado, R. Soil, Water and Nitrates Management in Horticultural Production. *Horticulturae* **2022**, *8*, 397. https://doi.org/10.3390/horticulturae8050397

Received: 26 April 2022
Accepted: 27 April 2022
Published: 1 May 2022

Publisher's Note: MDPI stays neutral with regard to jurisdictional claims in published maps and institutional affiliations.

The goal of this Special Issue, entitled "Soil, Water and Nitrates Management in Horticultural Production", is to examine recent advances in horticultural practices and strategies that can contribute to maintaining or increasing soil fertility and the efficiency of water and nitrogen use. The decrease in soil fertility, and in the quality and availability of irrigation water, is a reality that is maximized by global warming. On the other hand, nitrogen fertilization may contribute to nitrate leaching and/or the release of gases (carbon dioxide and nitrous oxide), with a great impact on global warming. Therefore, the management of soil, water, and nitrogen is critical to dealing with soil and water degradation, and with climate change. Moreover, such management has a decisive impact on plant growth and the quality of horticultural crops. In this Special Issue, seven scientific contributions were collected that can contribute to an increase in the use of these more sustainable factors. Three papers analyzed the use of organic waste materials as a way to reduce inorganic fertilization, improve soil fertility and plant growth, and promote a circular economy.

Jabborova et al. [1] reported that the addition of biochar to soil increased seed germination, total root length, plant growth, plant biomass, and the chlorophyll and carotenoid content of the ginger. Moreover, it was noted that it may improve soil nutrient availability, since it can increase the activity of soil enzymes.

Machado et al. [2] showed that the addition of municipal solid waste compost to soil, collected selectively, is a way to decrease inorganic fertilization, increase soil organic matter, and correct soil pH. The supplementation of municipal solid waste compost—collected selectively with reduced amounts of inorganic nitrogen, and applied using weekly fertigation—greatly increased spinach-shoot dry-weight and fresh yield. The addition of municipal solid waste compost to acid soil increased soil pH to values adequate for plant growth. The combined application of compost and inorganic nitrogen reduced inorganic N application, and replaced inorganic P and K fertilization to a significant extent.

Ronga et al. [3] analyzed the potential of an innovative process to produce controlled-release fertilizers using spent coffee grounds and biochar residues as raw materials to produce clay ceramic materials, in particular, lightweight aggregates. They showed that the lightweight clay ceramic aggregates containing spent coffee grounds, glass, and N fertilizers can be used for nursery grapevine production, reducing inorganic fertilization.

Badr et al. [4] analyzed the interaction of varying water-stress levels and arbuscular mycorrhizal fungi on the vegetative growth, reproductive behavior, fruit yield, water status, nutrient uptake, and soil fertility of field-grown eggplant in an arid region. They reported that the inoculation of mycorrhizal colonization increased the yield and regulated the physiological status of eggplant, associated with higher nutrient uptake at different water-stress levels. The mycorrhizal response showed better performance under severe water stress than under full irrigation conditions.

Di Mola, et al. [5] evaluated the possible effects of two mulching films (black polyethylene and brown photoselective film) and the use of a plant-growth-promoting product, containing *Trichoderma* spp., on the productive and qualitative traits of lettuce grown under four regimes of nitrogen (0, 30, 60 and 90 kg ha^{-1}). They concluded that the brown photoselective film increased the marketable yield of lettuce over black polyethylene. *Trichoderma* spp. did not affect the marketable yield, probably due to the short crop cycle. However, it had a positive effect on some quality traits, which, according to the authors, is an interesting starting point for further research. The nitrogen addition increased the marketable yield by up to 60 kg ha^{-1}, and improved chlorophyllous pigment biosynthesis, as well as antioxidant activities (lipophilic and hydrophilic) and bioactive compounds (phenols and total ascorbic acid).

The last two studies evaluated the influence of soil-water deficit on total sugar content in Meiwa kumquat trees (Iwasaki et al. [6]), and the potential of purslane for the phytoremediation of soils contaminated with Cr(VI) (Thalassinos et al.).

Iwasaki et al. [6] studied the influence of soil-water deficit on the accumulation of soluble sugars in the organs of Meiwa kumquat trees, and the relationship between the sugar content of each organ and the number of first-flush flowers. They reported that a soil-water deficit in Meiwa kumquat trees significantly increased the total sugar content of the xylem tissue of the scaffold branches to three times the value of the control.

Thalassinos et al. [7] analyzed the potential of purslane (Portulaca oleracea) as a phytoremediation species in Cr(VI)-contaminated soils. They concluded that purslane showed potential as a species to be successfully introduced to Cr(VI)-laden soils. Moreover, they also verified that the addition of N improved the plant's growth and physiological functions, even when exposed to high Cr(VI).

Conflicts of Interest: The authors declare no conflict of interest.

References

1. Jabborova, D.; Wirth, S.; Halwani, M.; Ibrahim, M.; Azab, I.; El-Mogy, M.; Elkelish, A. Growth Response of Ginger (*Zingiber officinale*). Its Physiological Properties and Soil Enzyme Activities after Biochar Application under Greenhouse Conditions. *Horticulturae* **2021**, *7*, 250. [CrossRef]
2. Machado, R.; Alves-Pereira, I.; Robalo, M.; Ferreira, R. Effects of Municipal Solid Waste Compost Supplemented with Inorganic Nitrogen on Physicochemical Soil Characteristics, Plant Growth, Nitrate Content, and Antioxidant Activity in Spinach. *Horticulturae* **2021**, *7*, 53. [CrossRef]
3. Ronga, D.; Parisi, M.; Barbieri, L.; Lancellotti, I.; Andreola, F.; Bignami, C. Valorization of Spent Coffee Grounds, Biochar and other residues to Produce Lightweight Clay Ceramic Aggregates Suitable for Nursery Grapevine Production. *Horticulturae* **2020**, *6*, 58. [CrossRef]
4. Badr, M.; El-Tohamy, W.; Abou-Hussein, S.; Gruda, N. Deficit Irrigation and Arbuscular Mycorrhiza as a Water-Saving Strategy for Eggplant Production. *Horticulturae* **2020**, *6*, 45. [CrossRef]
5. Di Mola, I.; Ottaiano, L.; Cozzolino, E.; Senatore, M.; Sacco, A.; El-Nakhel, C.; Rouphael, Y.; Mori, M. *Trichoderma* spp. and Mulching Films Differentially Boost Qualitative and Quantitative Aspects of Greenhouse Lettuce under Diverse N Conditions. *Horticulturae* **2020**, *6*, 55. [CrossRef]
6. Iwasaki, N.; Tamura, A.; Hori, K. Altered Carbohydrate Allocation Due to Soil Water Deficit Affects Summertime Flowering in Meiwa Kumquat Trees. *Horticulturae* **2020**, *6*, 49. [CrossRef]
7. Thalassinos, G.; Nastou, E.; Petropoulos, S.; Antoniadis, V. Nitrogen Effect on Growth-Related Parameters and Evaluation of Portulaca oleracea as a Phytoremediation Species in a Cr(VI)-Spiked Soil. *Horticulturae* **2021**, *7*, 192. [CrossRef]

 *horticulturae*

Article

Deficit Irrigation and Arbuscular Mycorrhiza as a Water-Saving Strategy for Eggplant Production

M. A. Badr [1,*], W. A. El-Tohamy [2], S. D. Abou-Hussein [2] and N. S. Gruda [3]

[1] Plant Nutrition Department, National Research Centre, Giza 12622, Egypt

[2] Vegetable Research Department, National Research Centre, Giza 12622, Egypt; waeleltohamy27@gmail.com (W.A.E.-T.); shaban_abouhussein@yahoo.com (S.D.A.-H.)

[3] Institute of Plant Sciences and Resource Conservation, Division of Horticultural Sciences, University of Bonn, 53111 Bonn, Germany; ngruda@uni-bonn.de

[*] Correspondence: badrmab@hotmail.com; Tel.: +20-122-4320-075; Fax: +20-233-370-931

Received: 15 July 2020; Accepted: 6 August 2020; Published: 11 August 2020

Abstract: Crop production in arid regions requires continuous irrigation to fulfill water demand throughout the growing season. Agronomic measures, such as roots-soil microorganisms, including arbuscular mycorrhizal (AM) fungi, have emerged in recent years to overcome soil constraints and improve water use efficiency (WUE). Eggplant plants were exposed to varying water stress under inoculated (AM+) and non-inoculated (AM−) to evaluate yield performance along with plant physiological status. Plants grown under full irrigation resulted in the highest fruit yield, and there were significant reductions in total yield and yield components when applying less water. The decline in fruit yield was due to the reduction in the number of fruits rather than the weight of the fruit per plant. AM+ plants showed more favorable growth conditions, which translated into better crop yield, total dry biomass, and number of fruits under all irrigation treatments. The fruit yield did not differ between full irrigation and 80% evapotranspiration (ET) restoration with AM+, but a 20% reduction in irrigation water was achieved. Water use efficiency (WUE) was negatively affected by deficit irrigation, particularly at 40% ET, when the water deficit severely depressed fruit yield. Yield response factor (Ky) showed a lower tolerance with a value higher than 1, with a persistent drop in WUE suggesting a lower tolerance to water deficits. The (Ky) factor was relatively lower with AM+ than with AM− for the total fruit yield and dry biomass (Kss), indicating that AM may enhance the drought tolerance of the crop. Plants with AM+ had a higher uptake of N and P in shoots and fruits, higher stomatal conductance (g_s), and higher photosynthetic rates (Pn), regardless of drought severity. Soil with AM+ had higher extractable N, P, and organic carbon (OC), indicating an improvement of the fertility status in coping with a limited water supply.

Keywords: drip irrigation; arbuscular mycorrhizal fungi; water relations; N and P status; soil organic carbon

---

1. Introduction

The search for additional water resources and improving the effective use of available water is an imperative strategy to overcome rising energy costs and water shortages in many parts of the world. This approach is challenging, especially in arid and semi-arid regions, which highlights an urgent solution for innovative irrigation strategies and optimized irrigation water management. Food production in the future needs efforts to understand the mechanisms of plant adaptation and tolerance of abiotic stress like water shortage, as these events are expected to intensify in the coming years [1]. Plants cope with drought stress by drought avoidance or drought tolerance, which include osmotic adjustment, the regulation of stomatal conductance and photosynthesis, and the regulation of water uptake and flow in their tissues [2]. Water-saving irrigation strategies, such as deficit irrigation (DI), may allow the optimization of water productivity, stabilizing the yield and improving the quality [3]. However, water productivity can be improved by developing DI strategies based on scientific principles while attempting to produce near-maximum yields even with a lesser water volume than that required to produce maximum productivity [4].

Strategies to sustain vegetative development and ecological succession against drought stress can be enhanced by associations between roots and soil microorganisms [5,6], water management [7,8], and arbuscular mycorrhizal (AM) fungi [9,10]. This association may include a suite of interrelated plant processes, primarily alleviating drought stress via direct water uptake and transport through fungal hyphae to the host plants [11] and enhanced nutrient uptake [10,12]. AM fungi can also change root hydraulic properties [13] that increase water supply to shoots and increase net photosynthetic rates under water stress conditions [14,15]. AM plants often have higher stomatal conductance at lower soil moisture [16] and sometimes regulate stomatal closure [17] in ways that may optimize the response to soil moisture content.

Crop production right now depends mainly on providing essential plant nutrients in the form of mineral fertilizers, which become significant components of modern agriculture. Research priorities for agricultural science are currently based on developing alternative methods of sustaining plant nutrition with lower inputs of mineral fertilizers [18]. One such possibility is to replace mineral fertilizers by organic inputs and to supplement plants with specific root-associated microbes. AM can be integrated into sustainable soil management, reducing environmental problems by decreasing the use of chemical fertilizers [19,20]. Since P diffusion is severely limited in dry soil [21], AM contributions to plant P may be especially acute under drought [22] or non-drought conditions [23,24]. Enhanced P nutrition by AM fungi during times of water deficit has been proposed as an essential factor for upgrading host plant drought tolerance [25,26]. These beneficial functions may have great importance relative to climate change [27,28], particularly with water in areas of agricultural expansion, especially in arid and semi-arid regions.

Eggplant is an economically important vegetable crop, with 51.288 million tons produced worldwide. Egyptian production ranks third place worldwide, with 1.194 metric tons from 0.485 million ha, and produces 2.3% of the world production [29]. Irrigation water availability has been identified as one of the major limiting factors of eggplant productivity, especially during the hot and dry summer periods in arid regions. Eggplant is relatively susceptible to drought stress, as especially in the reproductive phase [30], the susceptible period to water stress is extended, and a water shortage negatively and markedly affects all yield components. Lovelli et al. [31] demonstrated that eggplant sensitivity to water stress was expressed in high marketable yield decrements and a drop in water productivity.

The objective of this study was to examine the interaction of varying water stress levels and arbuscular mycorrhizal fungus on vegetative growth, reproductive behavior, fruit yield, water status, nutrient uptake, and soil fertility of field-grown eggplant in an arid region. The main scientific hypothesis of this study was that the inclusion of AM fungi in the soil/plant system would increase crop yield under deficit irrigation, and thus result in higher water use efficiency.

2. Materials and Methods

2.1. Location and Growth Conditions

Field experiments were conducted at the Main Research Station, National Research Center located in Nubaria province west of the Nile Delta of Egypt during the summer growing season (March-June) of 2017 and 2018 using a drip irrigation system. The site is situated in an arid climate at an altitude of 24 m above mean sea level and is intersected by 30°30 N latitude and 30°20 E longitude. The area has hot and dry summer months with little or no rain during the entire year. The mean monthly evapotranspiration (ET) ranged from 3.7 to 6.7 mm in the respective cropping seasons. The climatic parameters were taken from the Nubaria meteorological station according to the official data from the

Egyptian Ministry of Agriculture. The detailed climatic parameters recorded in 2017 and 2018 from March to June during the growing season are summarized in Table 1.

Table 1. Average monthly maximum (T_{max}) and minimum (T_{min}) temperature, relative humidity, rainfall, evapotranspiration (ET_0), and wind speed during growing seasons.

Month	T_{max} (°C)	T_{min} (°C)	Relative Humidity (%)	Rain Fall (mm)	ET_0 (mm d^{-1})	Wind Speed (km h^{-1})
			2017			
March	23.6	10.2	46	7.4	3.8	11.67
April	28.3	12.5	37	3.9	4.7	11.30
May	32.2	15.9	38	4.5	5.9	10.56
June	34.5	18.7	42	0	6.7	10.56
			2018			
March	23.8	10.1	45	6.2	3.7	11.85
April	28.3	12.6	36	2.4	4.7	11.65
May	32.4	15.9	38	3.6	5.9	10.56
June	36.2	18.7	43	0	6.7	10.19

The soil of the experimental site is sandy in texture (*Entisol-Typic Torripsamments*), composed of 86.5% sand, 9.2% silt, and 4.3% clay, with an alkaline pH of 8.2, electrical conductivity of 0.85 dS/m, 1.5% $CaCO_3$, and 0.42% organic matter in the upper 0–20 cm soil layer. The available N, P, and K were 12, 3, and 38 mg kg^{-1} soil, respectively, before the initiation of the experiment. The average soil water content at field capacity from the surface soil layer down to a depth of 80 cm at 20-cm intervals was 9.8%, and the permanent wilting point for the corresponding depths was 4.6%.

2.2. Experimental Design and Treatments

The experiment was conducted in a factorial randomized block design and included four irrigation treatments (100%, 80%, 60%, and 40% of crop ET, or $ET_{1.0}$, $ET_{0.8}$, $ET_{0.6}$, and $ET_{0.4}$) as the main factor and two arbuscular mycorrhizal treatments defined as (AM–) and (AM+) as the sub-main factor with three replications in 3.0 m wide × 10.0 m long plots of each treatment. Transplants of eggplant hybrid 'Black Moon' were grown from sterilized seeds in the nursery greenhouse for eight weeks. Vermiculite-based mycorrhizal inoculum carrying arbuscular mycorrhizal fungus (*Glomus intraradices*) was applied in an eggplant nursery at a rate of 100 g m^{-2} just before sowing seeds. The inoculum was supplied by the Microbiology Lab of the Soil, Water and Environment Research Institute, Agriculture Research Center, Egypt. The AM was originally cultured in roots of maize, which carried the propagules (spores, infected roots, soil) for three months before the beginning of the experiment. Non-mycorrhizal plant nurseries were maintained separately. Eggplant roots were tested for mycorrhizal colonization at the end of the last four weeks after inoculation (28 days after sowing). After the establishment of the symbiosis, all plants were transferred to the main field and transplanted at a spacing of 40 cm between plants and 100 cm between rows (25,000 plants per hectare) in early March 2017 and 2018. The plants were arranged in north-south-oriented soil beds, which had previously received 20 t ha^{-1} of organic manure for all treatments. Before cultivation of eggplants, drip tubing (twin-wall GR, 15 mm inner diameter, 40 cm dripper spacing delivering 2.5 L h^{-1} at an operating pressure 100 kPa) were placed on the soil surface beside each plant row at the center of the soil beds. All treatments received the same amount of nitrogen 300 kg ha^{-1} (as ammonium nitrate), phosphorus at 120 kg ha^{-1} (as phosphoric acid), and potassium at 250 kg ha^{-1} (as potassium sulphate) for the season. The total amount of NPK at variable levels was injected directly into the mainline of the drip system in a water-soluble form using a venturi-tube injector. Nitrogen and potassium fertilizers were applied at 6-day intervals in 15 equal doses of N starting two weeks after planting and stopped 30 days before the end of the cropping period while P was injected weekly, from the first week after planting until the last week of April.

2.3. Estimation of Crop Water Requirement

Reference evapotranspiration (ET_0) was calculated daily using Penman–Monteith's semi-empirical formula [32]. The actual evapotranspiration was estimated by multiplying reference evapotranspiration with crop coefficient (Kc) values ($ETc = ET_0 \times Kc$) for different months based on crop growth stages. Eggplant is about a 130-day-duration crop, which may be divided into four stages, (initial, 30 days; developmental, 40 days; middle, 40 days; and fruit maturity, 20 days.

The crop coefficient during the crop season was 0.45, 0.75, 1.15, and 0.80 at the initial, developmental, middle, and fruit maturity stages, respectively [32]. The total actual volumes of irrigation water applied at full irrigation during the entire season were 480 mm and 494 for 2017 and 2018, respectively. During the initial stage of growth, plants were irrigated daily to encourage establishment, but after that irrigation frequency was every two days. The irrigation treatments started at 14 days after transplanting (DAT) and ended at 120 DAT, respectively (15 days before the last harvest).

Seasonal values of the yield response factor (Ky) were calculated for each experimental year as follows:

$$1 - \left(\frac{Ya}{Ym} \right) = Ky \left(1 - \frac{ETa}{ETm} \right), \tag{1}$$

where Ym (kg ha^{-1}) and Ya (kg ha^{-1}) are the maximum (that obtained from the fully irrigated treatment) and actual yield, respectively. ETm (mm ha^{-1}) and ETa (mm ha^{-1}) are the maximum (that obtained from the fully irrigated treatment) and actual ET, respectively, and Ky is the yield response factor, defined as the decrease in yield per unit decrease in ET [33]. According to the Ky calculation, Kss was calculated by Equation (1) replacing Ym with the maximum total dry biomass (SSm) and Ya with the actual total dry biomass (SSa) as follows:

$$1 - \left(\frac{SSa}{SSm} \right) = Kss \left(1 - \frac{ETa}{ETm} \right), \tag{2}$$

where Kss indicates the biomass response factor, which is the correlation between the relative total dry biomass loss and relative ET reduction.

2.4. Physiological Measurements and Sampling

Monitoring of the diurnal change of stomatal conductance (g_s) and photosynthetic rate (Pn) were measured on the youngest fully-expanded leaves from the top of the canopy in each replicate with a portable photosynthesis system (ADC Bio-Scientific, UK). Measurements were taken on the central section of the youngest fully expanded leaf between 12:00 and 14:00 h (when the temperature reached the maximum at midday) about 4 h before the end of the irrigation cycle at the flowering stage on two plants per replicate.

2.5. Measurements of Crop Parameters

The fruits were harvested five times when they had attained full size depending on the end of the physiological maturity stage. Total fruit yield was recorded during the harvest on at least 25 plants in a row in each treatment in all the replications and data were presented as tons per hectare. Fruit parameters, such as the weight of fresh fruits, the number of fruits, and the average weight of the fruit per plant, were estimated until the final harvest in each picking period. Total dry biomass was evaluated by harvesting three representative plants per replicate of each treatment at the final harvest and shoot, and fruit tissues were weighed in the field (fresh weight), and then subsamples were dried at 70 °C to a constant mass in a forced-air oven for subsequent nutrient analysis. Water use efficiency (WUE) was calculated from the total fruit yield (kg ha^{-1}) divided by the seasonal crop water applied for each irrigation treatment during the growing season and expressed as kg ha^{-1} mm^{-1}. Seasonal N and P uptake were derived from the whole plant sample data (see below), and the amount of uptake was calculated as the product of crop dry weight and the N concentrations in the plant tissues. Postharvest fertilizer recovery was calculated using the following equation:

$$F \text{ recovery} = \left(\frac{Ft - F0}{F}\right) \times 100, \tag{3}$$

where Ft equals the total nutrient uptake (shoots + fruits) under treatment, F0 equals the total nutrient uptake of the unfertilized treatment, and F equals the total amount of nutrient applied during the whole season. The average crop N and P uptake from the unfertilized field plots (F0) were 14 and 4 kg N and P ha^{-1}, respectively, for the whole growing season.

2.6. Plant and Soil Analysis

Total N and P contents of the shoot and fruit tissues were determined for each year at the Analytical Research Lab (National Research Center, Cairo, Egypt) using standard methods of analysis. Tissue samples were ground to pass through a 0.5-mm screen and stored for dry weight analysis, with a thoroughly mixed 5-g portion of each sample stored. Tissue material was digested using H_2SO_4 in the presence of H_2O_2 and analyzed for total Kjeldahl N [34] and P by colorimetric ascorbic acid methodology [35]. In both years, soil samples were taken to determine the initial (0–40 cm) and the residual amount of ammonium (NH_4–N) and nitrate (NO^{-3}-N) at five soil core samples (10 cm apart) per plot, directly under drip lines. Ammonia and nitrate contents in each soil sample were extracted with $2M$ KCl, according to Jones and Case [34]. Microbial biomass carbon and 0.5 M K_2SO_4 extractable organic carbon (OC) were measured at all but the harvest sampling dates in surface soil (0–20 cm) by chloroform fumigation extraction followed by UV persulphate oxidation [36].

2.7. Statistical Analysis

Analysis of variance (ANOVA) was performed using the CoStat program (Version 6.311, CoHort, USA, 1998-2005) on the data of all yield and yield components, WUE, N and P uptake, and residual soil fertility measured in the experiments. Comparison of treatment means was carried out using the least significant difference (LSD) at $P < 0.05$. Linear regression analyses were made with Excel and Sigma Plot 9.0 (Systat Software, Inc., Point Richmond, CA, USA).

3. Results

3.1. Crop Biomass

Total fruit yield and total dry biomass for each irrigation treatment were similar in the two experimental years. Seasonal irrigation volume in the full irrigation treatment ($ET_{1.0}$) resulted in the maximum fruit yield and all yield traits under both AM treatments (Table 2). The reduction in irrigation water by 20% ($ET_{0.8}$) reduced the fruit yield by 22.9% and 21.5 % in 2017 and 2018 while the reduction by 40% ($ET_{0.6}$) and 60% ($ET_{0.4}$) reduced fruit yield by 44.1% and 71.0% and by 43.7% and 70.1% in both years, respectively, compared to the full irrigation treatment. Plants grown under mild water stress ($ET_{0.8}$) in AM+ had a fruit yield similar to plants receiving full irrigation in AM−, but a provision of 20% ($ET_{0.8}$) water saving was achieved during the entire season. AM+ plants averaged 27.8% and 27.5% higher fruit yield across irrigation treatments than AM− plants in both years, respectively.

Table 2. Fruit yield, total dry biomass, fruit number, and fruit weight of eggplant subjected to varying water stress and AM inoculation.

Year	AM [z]	Irrigation Levels				
		$ET_{1.0}$	$ET_{0.8}$	$ET_{0.6}$	$ET_{0.4}$	Mean
2017		Fruit yield (t ha^{-1})				
	AM−	42.75 [b,y]	32.82 [c]	23.75 [d]	12.32 [e]	27.91 [B]
	AM+	54.25 [a]	41.97 [b]	30.52 [c]	15.94 [e]	35.67 [A]
	Mean	48.50 [A]	37.40 [B]	27.14 [C]	14.13 [D]	
		Total dry biomass (t ha^{-1})				
	AM−	7.35 [c]	6.28 [d]	5.02 [e]	3.34 [g]	5.54 [B]
	AM+	9.33 [a]	8.03 [b]	6.68 [d]	4.32 [f]	7.09 [A]
	Mean	8.34 [A]	7.16 [B]	5.94 [C]	3.83 [D]	
		Fruit number (g plant^{-1})				
	AM−	23.64 [b]	17. 69 [c]	14.10 [d]	7.33 [e]	15.69 [B]
	AM+	28.46 [a]	23.90 [b]	18.97 [c]	9.09 [e]	20.10 [A]
	Mean	26.05 [A]	20.80 [B]	16.53 [C]	8.21 [D]	
		Fruit weight (g plant^{-1})				
	AM−	72.33 [a,b]	74.20 [a,b]	67.39 [a,b]	67.20 [a,b]	70.28 [A]
	AM+	76.26 [a]	70.23 [a,b]	64.35 [b]	70.18 [a,b]	70.25 [A]
	Mean	74.29 [A]	72.22 [A]	65.87 [B]	68.69 [B]	
2018		Fruit yield (t ha^{-1})				
	AM−	45.12 [b]	35.18 [c]	25.12 [d]	13.32 [e]	29.69 [B]
	AM+	57.32 [a]	44.27 [b]	32.54 [c]	17.38 [e]	37.88 [A]
	Mean	51.22 [A]	40.23 [B]	28.83 [C]	15.35 [D]	
		Total dry biomass (t ha^{-1})				
	AM−	7.76 [b,c]	6.73 [d]	5.50 [e]	3.61 [g]	5.90 [B]
	AM+	9.86 [a]	8.47 [b]	7.12 [c,d]	4.71 [f]	7.54 [A]
	Mean	8.81 [A]	7.60 [B]	6.31 [C]	4.16 [D]	
		Fruit number (g plant^{-1})				
	AM−	24.08 [b]	19.18 [c]	13.36 [d]	7.36 [e]	16.00 [B]
	AM+	29.23 [a]	25.51 [b]	18.52 [c]	9.39 [e]	20.66 [A]
	Mean	26.66 [A]	22.53 [B]	15.94 [C]	8.38 [D]	
		Fruit weight (g plant^{-1})				
	AM−	74.94 [a,b]	73.37 [a,b]	75.22 [a,b]	72.35 [a,b]	73.97 [A]
	AM+	78.44 [a]	69.41 [b]	70.27 [b]	74.07 [a,b]	73.05 [A]
	Mean	76.69 [A]	71.39 [B]	72.75 [B]	73.21 [B]	

[z] Abbreviations: AM = arbuscular mycorrhizal treatment; ET = evapotranspiration. [y] Values are significantly different based on the least significant difference (LSD) at $P \leq 0.05$. Capital letters indicate significant differences among main effect means, and small letters indicate significant differences within factorial treatments.

Moreover, the effect of AM fungi was more pronounced in the severe drought treatment than at full irrigation (yield was 29.3% and 30.5% higher in AM+ than AM- at $ET_{0.4}$ compared to 26.9% and 27.1% at $ET_{1.0}$ during the two seasons, respectively). Total dry biomass was proportional to total fruit yield production ($R^2 = 0.99$; $P = 0.05$), demonstrating that the crop regulates its fruit biomass relative to the total cumulative biomass. The total dry biomass showed more evident differences between full irrigation and water deficits at $ET_{0.6}$ (28.8%) and $ET_{0.4}$ (54.1%) compared to the $ET_{0.8}$ (14.2%) treatment. AM+ plants averaged 28.0% and 27.8% higher dry biomass across irrigation treatments than with AM− in both years, respectively.

The number of fruits per plant decreased with a decrease in the amount of water applied under both AM treatments, and the trend was similar to the fruit yield. The fruit weight was more resistant to a water deficit than the fruit number per plant, as the decrease in fruit yield in the water deficit treatments was mainly due to the decrease in the fruit number. The correlation analysis between the fruit yield and fruit number showed a very strong correlation ($R^2 = 0.987$; $P = 0.05$), which indicates that the increase in fruit yield was attributed mainly to the increase in fruit number (Figure 1a). On the other hand, the correlation analysis between the fruit yield and average fruit weight was weak ($R^2 = 0.236$;

$P = 0.05$), indicating that fruit weight was not the main reason behind yield reduction (Figure 1b). The relationship between fruit weight and the number of fruits per plant were not related ($R^2 = 0.165$; $P = 0.05$), where the decrease in the weight of fruits did not affect the number of fruits per plant. The number of fruits increased with AM+ under all water deficit treatments, and the trend was similar to the yield trend. The average fruit weight was not significantly affected by a water deficit up to $ET_{0.8}$ in the first season while it was only significant at $ET_{1.0}$ in the second season. Inoculation with AM did not show a significant effect on fruit weight under any deficit irrigation treatment. The interaction between irrigation and AM treatments for total yield and yield traits was significant, indicating a combined action between both treatments, although this effect was not always significant.

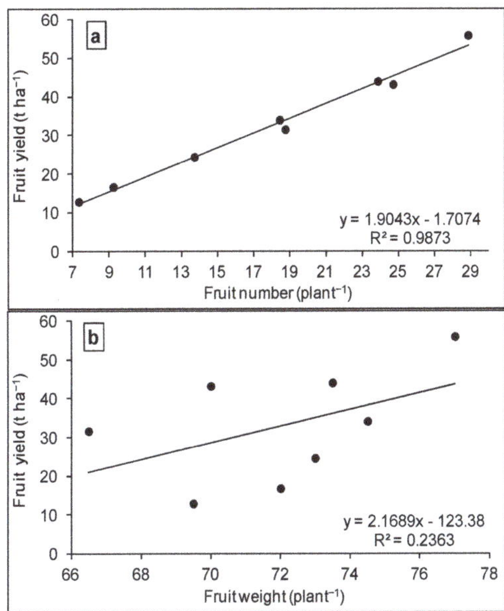

Figure 1. Relationship between the total fruit yield and fruit number per plant (**a**), and total fruit yield and average fruit weight per plant (**b**), of eggplant subjected to varying water stress and AM inoculation.

3.2. WUE and Production Function

The WUE was highest under full irrigation with a decreasing trend by deficit irrigation treatments (Table 3). However, the variation between WUE at the different irrigation treatments was not significant except for the $ET_{0.4}$ treatment, in which water stress severely depressed fruit yield by about 70% compared to full irrigation. Across different irrigation treatments, WUE increased by 28.9% and 27.5% with AM+ over AM− in the first and second seasons, respectively. This result may be because mycorrhizal inoculation has positive implications for eggplant culture in arid regions when water is not reasonably sufficient for plant growth.

Table 3. Water use efficiency (WUE - kg ha^{-1} mm^{-1}) of eggplant subjected to varying water stress and AM inoculation.

Year	AM ^{z}	Irrigation Levels				
		ET$_{1.0}$	ET$_{0.8}$	ET$_{0.6}$	ET$_{0.4}$	Mean
2017	AM−	86 ^{b,y}	82 ^{b}	78 ^{b}	59 ^{c}	76 ^{B}
	AM+	109 ^{a}	105 ^{a}	100 ^{a}	77 ^{b}	98 ^{A}
	Mean	98 ^{A}	93 ^{A}	89 ^{A}	68 ^{B}	
2018	AM−	89b ^{b}	86 ^{b}	81 ^{b}	63 ^{c}	80 ^{B}
	AM+	113 ^{a}	108 ^{a}	105 ^{a}	83 ^{b}	102 ^{A}
	Mean	101 ^{A}	97 ^{A}	93 ^{A}	73 ^{B}	

^{z} Abbreviations: AM = arbuscular mycorrhizal treatment; ET = evapotranspiration. ^{y} Values followed by different letters are significantly different based on the least significant difference (LSD) at $P \leq 0.05$. Capital letters indicate significant differences among main effect means, and small letters indicate significant differences within factorial treatments.

The regression equation fit for seasonal water use (ET) versus fruit yields showed that the increase in ET would induce a different response on yield (Figure 2). Fruit yields increased linearly with the total amount of water applied between planting and harvest over the range 208–496 mm in 2016 and 210–507mm in 2017, where single lines represented an average of AM− or AM+ during the two seasons. Over the range of water inputs, the yield in AM− increased by 107 kg ha^{-1} for each mm of water applied while the corresponding value in AM+ increased by 130 kg ha^{-1} mm for the same years. Both functions were significant and had a high coefficient of determination (R^2 = 0.998 and 0.997; P = 0.05) for the same seasons, respectively. The intercept on the x-axis of this relationship can be used as a measure of the amount of water lost by evaporation from the soil and canopy before crop establishment. However, this remaining water was relatively little with AM+, indicating that mycorrhiza can provide primarily support in the early stages of the plant of growth.

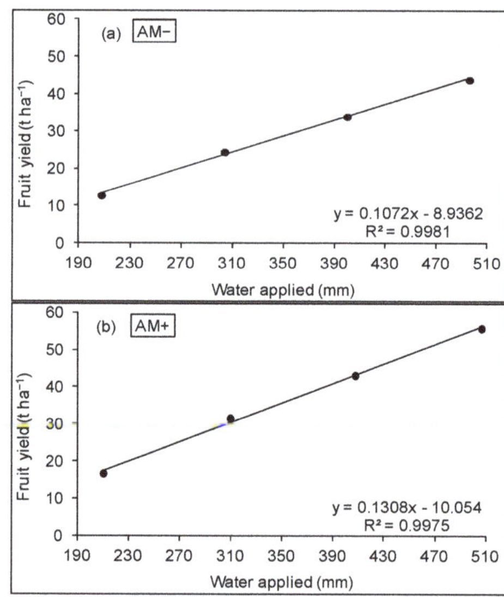

Figure 2. Relationship of total fruit yield versus the amounts of applied water (pooled AM− (**a**) and AM+ (**b**) of the two years), through regression analysis.

The yield response factor (Ky) indicates the level of tolerance of a crop to water deficits (the higher the value, the lower the tolerance). The yield response factor (Ky) was calculated for both the total yield (Ky) and total dry biomass (Kss) produced by the crop, considering the data of the two years

(Figure 3). The slope of the fitted regressions, which represents *Ky* and *Kss* factors, was 1.196 and 0.863 for fruit yield and total dry biomass in AM−and 1.164 and 0.836 for the same parameters in AM+, respectively, where in both cases, the reduction in crop productivity was proportionally higher than the relative ET deficit. However, the lower *Ky* was calculated for both parameters, indicating that eggplant with AM+ was relatively less affected by the soil water deficit than with AM−, which may help plants to cope with low water under field conditions. The calculation of this coefficient with respect to the plant total dry biomass rather than yield resulted in values < 1 (0.836 for AM−and 0.863 for AM+), which would seem to be contrary to fruit yield. The evidence supporting this result was related to the effect of irrigated treatments on the assimilate partitioning distribution (harvest index), which significantly changed among the irrigated treatments, ranging between 0.47 and 0.41.

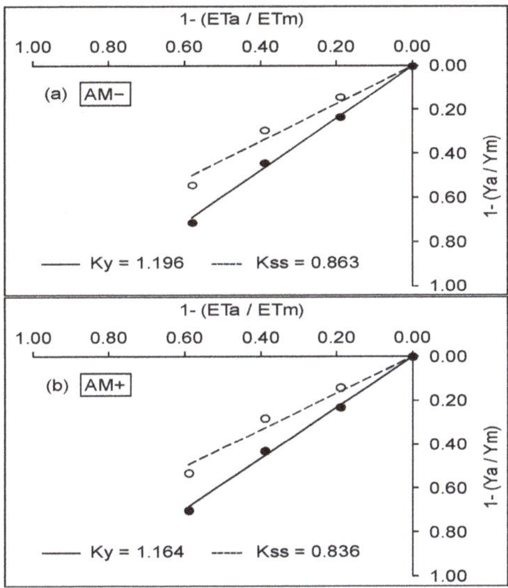

Figure 3. Relative yield decrease and total dry biomass relative decrease as a function of the relative ET decrease, measured in eggplant subjected to varying water stress and AM inoculation (pooled AM− (**a**) and AM+ (**b**) of the two years).

3.3. Plant Physiological Response

Plant physiological status showed remarkable differences between full irrigation and water deficit treatments during the flowering stage (Figure 4a,b). Considering all measurements taken when the temperature rose at midday, g_s and Pn in $ET_{1.0}$ plants averaged higher values than in water deficit treatments under both AM treatments. In contrast, the values in the mild water deficit $ET_{0.8}$ with AM+ were a little bit lower than other treatments. The values of g_s showed an appreciable decline by 29% and 40% in water deficit treatments at $ET_{0.6}$ and $ET_{0.4}$ with AM− vs. only 17% and 26% with the same treatments in AM+, respectively. Similarly, Pn values in $ET_{0.8}$ were unaffected at $ET_{0.8}$ by either AM treatment. At the same time, AM− contributed to a 25% and 31% decline at $ET_{0.6}$ and $ET_{0.4}$ with AM− vs. 16% and 22% only with AM+, respectively, than those obtained at full irrigation.

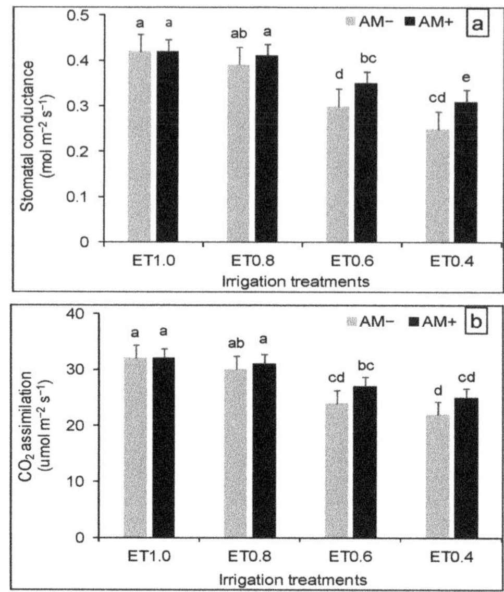

Figure 4. Changes of stomatal conductance (**a**), and CO_2 assimilation rate (**b**) at the flowering stage in eggplant subjected to varying water stress and AM inoculation. Data points are two years of pooled data of three replications.

3.4. Plant N and P Uptake

Substantial significant differences were found in total N and P uptake between various irrigation and AM treatments and such a trend of response was similar to the yield trend (Table 4). Total aboveground N uptake (shoots + fruits) was relatively higher in AM+ than in AM− plants under all irrigation treatments, resulting from higher dry biomass and N contents in the shoots and fruits (data not shown). Similarly, total aboveground P uptake was higher in AM+ plants, which corresponded to both a relatively higher plant P concentration and higher dry biomass. As a proportion of the N and P uptake, AM+ gave the most satisfactory recovery of both nutrients despite the sandy nature of the soil. This corresponded to a lower N:P ratio in AM+ plants mainly in the full irrigation treatment, corresponding to a relatively higher P uptake than N uptake, indicating a specific effect of AM+ on the absorption of phosphorus.

Table 4. Total N and P uptake in fruits and shoots of eggplant subjected to varying water stress and AM inoculation.

Year	Treatment [z]	N Uptake (kg ha^{-1})			N Recovery	P Uptake (kg ha^{-1})			P Recovery	N:P
		Fruit	Shoot	Total	%	Fruit	Shoot	Total	%	Ratio
2017	ET$_{1.0}$ AM−	75 [b,y]	69 [c]	144 [c]	44	12.0 [c]	9.8 [c]	21.8 [c]	16	6.6
	AM+	95 [a]	87 [a]	183 [a]	57	17.4 [a]	15.0 [a]	32.3 [a]	24	5.7
	ET$_{0.8}$ AM−	63 [c]	60 [d]	123 [d]	37	10.0 [d]	7.9 [d]	17.9 [d]	12	6.9
	AM+	80 [b]	77 [b]	157 [b]	48	14.6 [b]	12.3 [b]	26.9 [b]	20	5.8
	ET$_{0.6}$ AM−	47 [d]	48 [e]	95 [e]	28	7.0 [e]	6.3 [e]	13.3 [e]	9	7.1
	AM+	60 [c]	66 [c,d]	126 [d]	38	9.8 [d]	8.9 [c]	18.7 [c,d]	13	6.7
	ET$_{0.4}$ AM−	30 [f]	31 [f]	61 [g]	16	4.2 [f]	4.1 [f]	8.3 [f]	4	7.4
	AM+	38 [e]	43 [e]	82 [f]	23	6.3 [e]	5.8 [e]	12.1 [e]	8	6.7

Table 4. *Cont.*

Year	Treatment [z]	N Uptake (kg ha^{-1})			N Recovery %	P Uptake (kg ha^{-1})			P Recovery %	N:P Ratio
		Fruit	Shoot	Total		Fruit	Shoot	Total		
2018	ET$_{1.0}$ AM−	78 [b]	73 [c]	150 [c]	46	12.6 [c]	10.4 [d]	23.0 [c]	17	6.5
	AM+	99 [a]	92 [a]	191 [a]	60	18.3 [a]	15.8 [a]	34.2 [a]	26	5.6
	ET$_{0.8}$ AM−	67 [c]	64 [c]	132 [d]	40	10.7 [d]	9.2 [c]	19.9 [d]	14	6.6
	AM+	85 [b]	81 [b]	166 [b]	51	15.4 [b]	13.9 [b]	29.3 [b]	22	5.7
	ET$_{0.6}$ AM−	50 [d]	51 [d]	101 [e]	30	7.4 [e]	6.4 [d]	13.8 [e]	9	7.3
	AM+	64 [c]	70 [c]	135 [d]	41	10.5 [d]	10.3 [c]	21.7 [c,d]	15	6.5
	ET$_{0.4}$ AM−	32 [f]	34 [e]	66 [f]	18	4.5 [f]	4.2 [e]	8.7 [f]	5	7.6
	AM+	42 [e]	47 [d]	89 [e]	26	6.8 [e]	6.9 [d]	13.7 [e]	9	6.5

[z] Abbreviations: AM = arbuscular mycorrhizal treatment; ET = evapotranspiration. [y] Values within the column followed by different letters are significantly different based on the least significant difference (LSD) at $P \leq 0.05$.

3.5. Soil Fertility

The analysis of variance within each year revealed no significant effect for irrigation treatments alone, or when combined with AM+ on soil fertility; therefore, only the latter factor is only discussed. Before plant cultivation, initial OC in the soil was 0.25% while at harvest of the first and second season, there was a trend toward higher OC in soil with AM+ plants, reaching 26% and 31% over AM− in the first and the second seasons, respectively (Table 5). AM+ within irrigation treatments led to varying amounts of available NPK in the soil after harvest. The available N and K content with AM+ increased by 20 - 25% and 15 - 17% over AM− in the same season, respectively, while P content increased markedly by 30-50% regardless of drought stress.

Table 5. Residual soil fertility after the harvest of eggplant subjected to varying water stress and AM inoculation.

Year	Treatments [z]	Organic C (%)	Available Nutrients (mg kg^{-1} soil)		
			N	P	K
2017	AM−	0.27 [a,y]	15 [a]	10 [a]	40 [a]
	AM+	0.34 [b]	18 [b]	13 [b]	46 [b]
2018	AM−	0.29 [a]	19 [a]	10 [a]	42 [a]
	AM+	0.38 [b]	24 [b]	15 [b]	49 [b]

[z] Abbreviations: AM = arbuscular mycorrhizal treatment. [y] Values within the column followed by different letters are significantly different based on the least significant difference (LSD) at $P \leq 0.05$.

4. Discussion

4.1. Yield Response and Plant Water Relations

The maximum yield traits under ET$_{1.0}$ were mainly due to the sufficient available water in the soil, which led to an increase in water and nutrient absorption and consequently in the metabolic mechanisms in the plants, increasing the fruit yield and total dry biomass. However, there were significant reductions in the total yield when less water was applied, and the negative response to the irrigation deficit was more evident when the water supply was less than 20% of the full irrigation demand. The shortage in the yield of many crops under lower available water may be due to a reduced soil water content, which has been shown to delay rooting [37], along with a reduction in leaf area, root system, and shoot biomass, and a low photosynthetic rate [38]. Fruit yield and total dry biomass with AM+ was higher during both the seasons across different irrigation regimes, with the total yield 28% higher than AM− plants.

The ability of AM fungi to protect the host plant against progressive drought appears to be related to the intrinsic capacity of mycorrhizal fungi to resist drought stress. Thus, the main effect of AM on plant growth was in fruit rather than shoots, which increased the total dry biomass. The association with AM fungi increased the eggplant yield and fruit numbers in both full and deficit irrigation treatments; a higher fruit set likely occurred in AM+ plants but without other substantial changes in fruit weight. The higher fruit biomass in AM+ plants and few differences in shoot biomass pointed to a specific effect on fruit rather than a general effect on plant growth. Eggplant showed a remarkable decrease in the harvest index (HI), which changed significantly among irrigation treatments (data not shown), indicating that water deficits had a substantial consequence on saleable production. However, AM fungi has been shown to affect plant reproductive growth [39,40], including increasing the total number of flowers as well as the proportion of flowers setting as fruit [41].

The correlation analysis between fruit yield and average fruit weight showed a weak correlation, indicating that the change in yield was not affected by fruit weight [42,43]. However, eggplant tends to abort some flowers when sensing water stress to limit the number of fruit reaching the maturity stage to as few as possible. Aujla et al. [24] and Colak et al. [43] reported similar results with a decrease in the fruit yield of eggplant at a reduced water level and the formation of fewer fruits. The fruit size was not affected.

The response to mycorrhizal colonization was relatively greater for drought-stressed than full irrigation conditions, which may be explained by a higher absorption root surface area or a significant proliferation of root or hydraulic differences between root systems [44,45]. Drought stress may impede the translocation of nutrients and metabolites from the leaf to the reproductive organs. The improved nutritional status and relative water content caused by mycorrhizal colonization would have alleviated drought effects and promoted fruit production with changing water stress. The experimental data agree with the findings of others that mycorrhizal aided host plant drought tolerance and the associated yield increase under water deficit conditions [9,25].

Higher rates of root sap exudation in AM+ plants may reflect the reason behind higher root osmotic hydraulic conductance, a pathway for water uptake that may play an essential role under dry conditions [26,46]. Irrespective of irrigation treatments, AM+ improved the plant water status, suggesting that AM+ roots could extract more water from deeper soil profiles or that AM+ plants regulated daily leaf gas exchange. These findings suggest that AM affected a suite of interrelated plant drought responses that together enabled plants to produce higher yields. Previous work from Cavagnaro et al. [47] showed similar microbial communities (AM) in the soil around roots, suggesting that these changes may be relatively minor. However, there is still a possibility that there are micro-scale fungal–bacterial interactions that affect nutrient availability and uptake by the plant.

Water use efficiency (WUE) is defined as the ratio between the actual yield achieved and the total water use, including rainfall, and is expressed in physical terms (kg ha mm^{-1}). WUE did not cause significant variation due to different deficit irrigation treatments except for ET$_{0.4}$ when the yield was severely depressed by water deficit (Table 3). Although, in a limited way, WUE showed a decreasing trend when the water supply was lower than 100% ET of the crop, indicating sensitivity to water stress [31]. However, AM fungi contributed consistently to higher WUE under all deficit irrigation treatments in both years, leading to a higher fruit yield. The regression equations' fit for crop ET versus fruit yields showed that the same increase in ET would induce a different improvement on eggplant yields for different irrigation treatments (Figure 2a,b). Significant linear relationships were established between the total applied water and fresh fruit yield as indicated by the slopes of the lines shown in Figure 2. The slope of the line for AM+ was steeper than for AM−, where the fruit yield increased by 130 t ha^{-1} for AM+ versus 107 t ha^{-1} for AM−, respectively.

The sensitivity of eggplant to water stress due to deficit irrigation, as expressed by (*Ky* and *Kss*), indicates the level of tolerance of a crop to water stress (the value is over than 1 when the yield declines proportionally to the ET deficit). Fruit yield and total dry biomass decreases by decreasing the water deficit are higher than proportional to the relative ET decrease. In this regard, eggplant seems to be more sensitive to a water deficit than other vegetable crops, e.g., tomato, where *Ky* was lower than one calculated in a Mediterranean region [48]. The calculation of *Ky* gave a value > 1 according to FAO, which means a particular sensitivity to a water deficit; in other words, a crop production decrease due to water deficit treatments, which is more than proportional to the ET decrease. However, higher values were obtained in AM+ for fruit yield and total dry biomass than in AM−, showing the importance of mycorrhizal inoculation in supporting water stress tolerance to stress-sensitive crops. The relatively lower *Ky* factor under AM+ could account for a guarantee of the higher water uptake capacity by the roots' association with stimulated soil OC, which in turn improves soil conditions for plant growth [26,49]. Thus, the water deficit imposed a substantial consequence on fresh fruit production that ranged between 22% and 70% due to water stress. In comparison, the measured total dry biomass decrements at the same water levels only ranged between 12% and 51%. Thus, in eggplant, the fresh fruit yield decrease was more proportional to the ET drop, and this explains the differences in terms of *Ky* and *Kss* in this crop.

4.2. Photosynthesis, Nutrient Status, and Soil OC

A trend toward higher gs and Pn in AM+ plants under water stress was indicated by the assimilation of higher C for vegetative and fruit growth. In contrast, relative inhibition in the AM− plant was observed to avoid rapid loss of water through transpiration. However, the differences in gs in AM+ vs. AM− plants have been attributed to plant size and C dynamics over the whole growing season as a driver in water relations. AM fungi also regulate diurnal patterns of leaf gas exchange, for instance, by maximizing C gain through increased stomatal conductance early in the day when the vapor pressure is lower followed by a reduction in gs in the afternoon [50]. This could explain how AM+ plants could control higher gs when the daily air temperature peaks despite a larger canopy size.

Enhanced Pn in AM+ plants may result from higher gs, increased N and P nutrition, and/or higher C sink stimulation. Moreover, higher gs would increase CO_2 diffusion to sites of carboxylation and support higher Pn [26,51]. That M+ plants appeared to optimize responses to the soil moisture content in ways that would maximize growth agrees with studies in controlled environments that show AM+ plants respond more quickly than AM− plants to changes in soil moisture [17,26,52]. These results suggest that AM affects a range of biological processes associated with the plant drought response that produces a higher crop yield.

Water deficit plants had the lowest N and P uptake since the absorption of elements mainly depends on the available water in the soil. Total uptake of N was higher in AM+ plants than the corresponding values in AM− plants when considering the water treatments together, due to the higher fruit biomass in AM+ plants. As the greater demand for N and drought impedes the mobility of nitrate, mycorrhizal colonization is essential for the host plant N nutrition under water deficit conditions. The external mycelium of AM can transport 40% of the added N under moderate drought conditions, which modifies the N acquisition and assimilation by host plant roots [53]. An enhanced nutritional and water status of mycorrhizal plants assists in the production of a higher number of flowers and fruits and a greater degree of conversion of flowers into fruits, where abortion of flowers is a significant constraint that leads to lower productivity in many plants [25,43]. Total P uptake was also higher in AM+ plants due to the relatively higher plant P uptake than N uptake in these plants. The remarkable increases in plant N and P uptake observed in AM+ plants may have affected growth, especially fruit production [54] and the physiological status (photosynthetic rate and stomatal conductance [26,55]. The enhanced capacity for P uptake by AM+ fungi was expected to be more effective under drought conditions as AM could transfer extra P to the root system by passing direct uptake [56,57]. This contribution resulted in a better N:P ratio in fruit and total dry biomass but mainly in the full irrigation treatment. Higher uptake of N and P in AM+ plants can contribute to a significant increase in fruit biomass, where a large proportion of nutrients will be stored in fruits that will be reflected in plant production.

The organic C and available N and P content increased relatively with AM+ plants at the harvest of both seasons as a result of organic substances and microbial activity in the soil (Table 5). The increase in available P could be attributed to the mineralization of organic P, solubilization action of certain organic acids, and displacement of phosphate with organic anions. The equivalent quantity of these nutrients from such an organic source will be available in a fair manner and could be an additive source for plant nutrition for sustaining soil fertility buildup.

5. Conclusions

The inoculation of mycorrhizal colonization increased the yield and regulated the physiological status of eggplant associated with a higher nutrient uptake and soil fertility at different water stress, and potentially managed water drought tolerance. The mycorrhizal response showed a better performance under severe water stress than full irrigation conditions. This proved that mycorrhizal fungi play an essential role in plant responses under adverse plant and soil conditions, particularly in sandy soil. Strategies that boost water-saving management in agriculture systems may be suggested to include services provided by mycorrhizal associations from detrimental effects of drought stress in arid and semi-arid regions.

Author Contributions: M.A.B. and W.A.E.-T. conceptualized and designed the experiment, M.A.B, W.A.E.-T. and S.D.A.-H. performed the experiments, M.A.B. and W.A.E.-T. analyzed the data and wrote the paper; N.S.G. revised and finalized the paper. All authors have read and agreed to the published version of the manuscript.

Funding: This research received no external funding.

Conflicts of Interest: The authors declare no conflict of interest.

Abbreviations

AM	arbuscular mycorrhizal;
AMF	arbuscular mycorrhizal fungi;
AM+	inoculated;
AM−	non-inoculated;
WUE	water use efficiency;
ET	evopotranspiration;
ET_0	evapotranspiration;
Tmax	average monthly maximum temperature;
Tmin	and average monthly minimum temperature;
Ym	reference maximum yield;
Ya [1]	actual yield;
Kss	biomass response factor;
Ky	yield response factor;
SS	total dry biomass;
g_s	stomatal conductance;
Pn	photosynthetic rate;
Ft	total nutrient uptake;
F0	total nutrient uptake under unfertilized treatment;
F	total amount of nutrient applied during the whole season;
DAT	days after transplantation;
HI	harvest index;
OC	organic carbon.

References

1. Elliott, J.; Deryng, D.; Müller, C.; Frieler, K.; Konzmann, M.; Gerten, D.; Glotter, M.; Flörke, M.; Wada, Y.; Best, N.; et al. Constraints and potentials of future irrigation water availability on agricultural production under climate change. *Proc. Natl. Acad. Sci. USA* **2014**, *111*, 3239–3244. [CrossRef] [PubMed]
2. Ruiz-Lozano, J.M.; Porcel, R.; Azcón, R.; Bárzana, G.; Aroca, R. Contribution of arbuscular mycorrhizal symbiosis to plant drought tolerance: State of the art. In *Responses to Drought Stress: From Morphological to Molecular Features*; Aroca, R., Ed.; Springer: Berlin/Heidelberg, Germany, 2012; pp. 335–362.
3. Costa, J.M.; Ortuño, M.F.; Chaves, M.M. Deficit irrigation as a strategy to save water: Physiology and potential application to horticulture. *J. Integr. Plant Biol.* **2007**, *49*, 1421–1434. [CrossRef]

4. Badr, M.A.; El-Tohamy, W.A.; Zaghloul, A.M. Yield and water use efficiency of potato grown under different irrigation and nitrogen levels in an arid region. *Agric. Water Manag.* **2012**, *110*, 9–15. [CrossRef]

5. Bardgett, R.D.; van der Putten, W.H. Belowground biodiversity and ecosystem functioning. *Nature* **2014**, *515*, 505–511. [CrossRef]

6. Mohan, J.E.; Cowden, C.C.; Baas, P.; Dawadi, A.; Frankson, P.T.; Helmick, K.; Hughes, E.; Khan, S.; Lang, A.; Machmuller, M.; et al. Mycorrhizal fungi mediation of terrestrial ecosystem responses to global change: Mini-review. *Fungal Ecol.* **2014**, *10*, 3–19. [CrossRef]

7. Badr, M.A.; Abou Hussein, S.D.; El-Tohamy, W.A.; Gruda, N. Efficiency of subsurface drip irrigation for potato production under different dry stress conditions. *Healthy Plants* **2010**, *62*, 63–70. [CrossRef]

8. Badr, M.A.; El-Tohamy, W.A.; Hussein, S.D.A.; Gruda, N. Tomato yield, physiological response, water and nitrogen use efficiency under deficit and partial root zone drying irrigation in an arid region. *J. Appl. Bot. Food Qual.* **2018**, *91*, 332–340. [CrossRef]

9. Augé, R. Water relations, drought and vesicular-arbuscular mycorrhizal symbiosis. *Mycorrhiza* **2001**, *11*, 3–42. [CrossRef]

10. Baum, C.; El-Tohamy, W.; Gruda, N. Increasing the productivity and product quality of vegetable crops using arbuscular mycorrhizal fungi: A review. *Sci. Hortic.* **2015**, *187*, 131–141. [CrossRef]

11. Augé, R.M.; Toler, H.D.; Moore, J.L.; Cho, K.; Saxton, A.M. Comparing contributions of soil versus root colonization to variations in stomatal behavior and soil drying in mycorrhizal Sorghum bicolor and Cucurbita pepo. *J. Plant Physiol.* **2007**, *164*, 1289–1299. [CrossRef]

12. Michalis, O.; Ioannides, I.M.; Ehaliotis, C. Mycorrhizal inoculation affects arbuscular mycorrhizal diversity in watermelon roots but leads to improved colonization and plant response under water stress only. *Appl. Soil Ecol.* **2013**, *63*, 112–119. [CrossRef]

13. Aroca, R.; Vernieri, P.; Ruiz-Lozano, J.M. Mycorrhizal and non-mycorrhizal Lactuca sativa plants exhibit contrasting responses to exogenous ABA during drought stress and recovery. *J. Exp. Bot.* **2008**, *59*, 2029–2041. [CrossRef]

14. Kaschuk, G.; Kuyper, T.W.; Leffelaar, P.A.; Hungria, M.; Giller, K.E. Are the rates of photosynthesis stimulated by the carbon sink strength of rhizobial and arbuscular mycorrhizal symbioses? *Soil Biol. Biochem.* **2009**, *41*, 1233–1244. [CrossRef]

15. Birhane, E.; Sterck, F.J.; Fetene, M.; Bongers, F.; Kuyper, T.W. Arbuscular mycorrhizal fungi enhance photosynthesis, water use efficiency, and growth of frankincense seedlings under pulsed water availability conditions. *Oecologia* **2012**, *169*, 895–904. [CrossRef] [PubMed]

16. Augé, R.M.; Toler, H.D.; Saxton, A.M. Arbuscular mycorrhizal symbiosis alters stomatal conductance of host plants more under drought than under amply watered conditions: A meta-analysis. *Mycorrhiza* **2015**, *25*, 13–24. [CrossRef]

17. Lazcano, C.; Barrios-Masias, F.H.; Jackson, L.E. Arbuscular mycorrhizal effects on plant water relations and soil greenhouse gas emissions under changing moisture regimes. *Soil Biol. Biochem.* **2014**, *74*, 184–192. [CrossRef]

18. Foley, J.A.; Ramankutty, N.; Brauman, K.A.; Cassidy, E.S.; Gerber, J.S.; Johnston, M.; Mueller, N.D.; O'Connell, C.; Ray, D.K.; West, P.C.; et al. Solutions for a cultivated planet. *Nature* **2011**, *478*, 337–342. [CrossRef]

19. Adesemoye, A.O.; Torbert, H.A.; Kloepper, J.W. Plant growth-promoting rhizobacteria allow reduced application rates of chemical fertilizers. *Microb. Ecol.* **2009**, *58*, 921–929. [CrossRef]

20. Hungria, M.; Nogueira, M.A.; Araujo, R.S. Co-inoculation of soybeans and common beans with rhizobia and azospirilla: Strategies to improve sustainability. *Biol. Fertil. Soils* **2013**, *49*, 791–801. [CrossRef]

21. Suriyagoda, L.D.B.; Ryan, M.H.; Renton, M.; Lambers, H. Plant responses to limited moisture and phosphorus availability: A meta-analysis. *Adv. Agron.* **2014**, *124*, 143–200.

22. Neumann, E.; George, E. Colonisation with the arbuscular mycorrhizal fungus Glomus mosseae (Nicol. & Gerd.) enhanced phosphorus uptake from dry soil in *Sorghum bicolor* (L.). *Plant Soil* **2004**, *261*, 245–255.

23. George, E.; Marschner, H.; Jakobsen, I. Role of arbuscular mycorrhizal fungi in uptake of phosphorus and nitrogen from soil. *Crit. Rev. Biotechnol.* **1995**, *15*, 257–270. [CrossRef]

24. Hetrick, B.A.D.; Wilson, G.W.T.; Todd, T.C. Mycorrhizal response in wheat cultivars: Relationship to phosphorus. *Can. J. Bot.* **1996**, *74*, 19–25. [CrossRef]

25. Subramanian, K.S.; Santhanakrishnan, P.; Balasubramanian, P. Responses of field grown tomato plants to arbuscular mycorrhizal fungal colonization under varying intensities of drought stress. *Sci. Hortic.* **2006**, *107*, 245–253. [CrossRef]

26. Bowles, T.M.; Barrios-Masias, F.H.; Carlisle, E.A.; Cavagnaro, T.R.; Jackson, L.E. Effects of arbuscular mycorrhizae on tomato yield, nutrient uptake, water relations, and soil carbon dynamics under deficit irrigation in field conditions. *Sci. Total Environ.* **2016**, *566*, 1223–1234. [CrossRef]

27. Gruda, N.; Bisbis, M.B.; Tanny, J. Impacts of protected vegetable cultivation on climate change and adaptation strategies for cleaner production—A review. *J. Clean. Prod.* **2019**, *225*, 324–339. [CrossRef]

28. Gruda, N.; Bisbis, M.B.; Tanny, J. Influence of climate change on protected cultivation: Impacts and sustainable adaptation strategies—A review. *J. Clean. Prod.* **2019**, *225*, 481–495. [CrossRef]

29. FAO. FAOSTAT. Food and Agriculture Organization of the United Nations. 2016. Available online: http://faostat.fao.org/statistics (accessed on 10 August 2020).

30. Karam, F.; Sabiha, R.; Skaf, S.; Breidy, J.; Rouphael, Y.; Balendonck, J. Yield and water use of eggplants (*Solanum Melongena* L.) under full and deficit irrigation regimes. *Agric. Water Manage.* **2011**, *98*, 1307–1316. [CrossRef]

31. Lovelli, S.; Perniola, M.; Ferrara, A.; Di Tommaso, T. Yield response factor to water (Ky) and water use efficiency of *Carthamus tinctorius* L. and *Solanum melongena* L. *Agric. Water Manag.* **2007**, *92*, 73–80. [CrossRef]

32. Allen, R.G.; Pereira, L.S.; Raes, D.; Smith, M. *Crop Evapotranspiration. Guidelines for Computing Crop Water Requirements. FAO Irrigation and Drainage. Paper No. 56*; FAO: Rome, Italy, 1998; p. 300.

33. Stewart, J.I.; Cuenca, R.H.; Pruit, W.O.; Hagan, R.M.; Tosso, J. *Determination and Utilization of Water Production Functions for Principal California Crops. W-67 California Contribution Project Report*; University of California: Davis, CA, USA, 1977.

34. Jones, J.B.J.; Case, V.W. *Sampling, Handling and Analyzing Plant Tissue Samples*; Westerman, R.L., Ed.; Soil Testing and Plant Analysis: Madison, WI, USA, 1991; pp. 289–427.

35. Eaton, A.D.; Clesceri, L.S.; Greenberg, A.E. *Standard Methods for the Examination of Water and Wastewater*, 19th ed.; American Public Health Association: Washington, DC, USA, 1995.

36. Wu, J.; Joergensen, R.G.; Pommerening, B.; Chaussod, R.; Brookes, P.C. Measurement of soil microbial biomass C by fumigation-extraction: An automated procedure. *Soil Biol. Biochem.* **1990**, *22*, 1167–1169. [CrossRef]

37. Bathke, G.R.; Gassel, D.K.; Hargrove, W.L.; Porter, P.M. Modification of soil physical properties and root growth response. *Soil Sci.* **1992**, *154*, 316–329. [CrossRef]

38. Masle, J.; Passiour, J.B. The effect of soil strength on the growth of young wheat plants. *Aust. J. Plant Physiol.* **1987**, *14*, 643–656. [CrossRef]

39. Bryla, D.R.; Koide, R.T. Regulation of reproduction in wild and cultivated *Lycopersicon esculentum* Mill. by vesicular-arbuscular mycorrhizal infection. *Oecologia* **1990**, *84*, 74–81. [CrossRef]

40. Poulton, J.L.; Bryla, D.; Koide, R.T.; Stephenson, A.G. Mycorrhizal infection and high soil phosphorus improve vegetative growth and the female and male functions in tomato. *New Phytol.* **2002**, *154*, 255–264. [CrossRef]

41. Conversa, G.; Lazzizera, C.; Bonasia, A.; Elia, A. Yield and phosphorus uptake of a processing tomato crop grown at different phosphorus levels in a calcareous soil as affected by mycorrhizal inoculation under field conditions. *Biol. Fertil. Soils* **2013**, *49*, 691–703. [CrossRef]

42. Aujla, M.S.; Thind, H.S.; Buttar, G.S. Fruit yield and water use efficiency of eggplant (*Solanum melongema* L.) as influenced by different quantities of nitrogen and water applied through drip and furrow irrigation. *Sci. Hortic.* **2007**, *112*, 142–148. [CrossRef]

43. Colak, Y.B.; Yazar, A.; Çolak, I.; Akça, H.; Duraktekin, G. Evaluation of crop water stress index (CWSI) for eggplant under varying irrigation regimes using surface and subsurface drip systems. *Agric. Sci. Procedia* **2015**, *4*, 372–382. [CrossRef]

44. Kothari, S.K.; Marschner, H.; George, E. Effect of VA fungi and rhizosphere organisms root and shoot morphology, growth and water relations in maize. *New Phytol.* **1990**, *116*, 303–311. [CrossRef]

45. Augé, R.M.; Duan, X.; Ebel, R.C.; Stodola, A.J.W. Nonhydraulic signaling of soil drying in mycorrhizal maize. *Planta* **1994**, *193*, 74–82. [CrossRef]

46. Barrios-Masias, F.H.; Knipfer, T.; McElrone, A. Differential responses of grapevine rootstocks to water stress are associated with adjustments in fine root hydraulic physiology and suberization. *J. Exp. Bot.* **2015**, *66*, 6069–6078. [CrossRef]

47. Cavagnaro, T.R.; Jackson, L.E.; Six, J.; Ferris, H.; Goyal, S.; Asami, D.; Scow, K.M. Arbuscular mycorrhizas, microbial communities, nutrient availability, and soil aggregates in organic tomato production. *Plant Soil* **2006**, *282*, 209–225. [CrossRef]

48. Patanè, C.; Tringali, S.; Sortinob, O. Effects of deficit irrigation on biomass, yield, water productivity and fruit quality of processing tomato under semi-arid Mediterranean climate conditions. *Sci. Hortic.* **2011**, *129*, 590–596. [CrossRef]

49. Cheng, W. Priming effect: Its functional relationships with microbial turnover, Rhizosphere evapotranspiration, and C–N budgets. *Soil Biol. Biochem.* **2009**, *41*, 1795–1801. [CrossRef]

50. Richards, R.A. Selectable traits to increase crop photosynthesis and yield of grain crops. *J. Exp. Bot.* **2000**, *51*, 447–458. [CrossRef]

51. Chitarra, W.; Pagliarani, C.; Maserti, B.; Lumini, E.; Siciliano, I.; Cascone, P.; Schubert, A.; Gambino, G.; Balestrini, R.; Guerrieri, E. Insights on the Impact of Arbuscular Mycorrhizal Symbiosis on Tomato Tolerance to Water Stress. *Plant Physiol.* **2016**, *171*, 1009–1023. [CrossRef]

52. Duan, X.; Neuman, D.; Reiber, J.; Green, C.; Saxton, A.; Augé, R. Mycorrhizal influence on hydraulic and hormonal factors implicated in the control of stomatal conductance during drought. *J. Exp. Bot.* **1996**, *47*, 1541–1550. [CrossRef]

53. Subramanian, K.S.; Charest, C. Acquisition of N by external hyphae of an arbuscular mycorrhizal fungus and its impact on physiological responses in maize under drought-stressed and well-watered conditions. *Mycorrhiza* **1999**, *9*, 69–75. [CrossRef]

54. Tei, F.; Benincasa, P.; Guiducci, M. Critical nitrogen concentration in processing tomato. *Eur. J. Agron.* **2002**, *18*, 45–55. [CrossRef]

55. Liu, C.; Ravnskov, S.; Liu, F.; Rubaek, G.H. Arbuscular mycorrhizal fungi alleviate abiotic stresses in potato plants caused by low phosphorus and deficit irrigation/partial root-zone drying. *J. Agric. Sci.* **2018**, *156*, 46–58. [CrossRef]

56. Cramer, M.; Hawkins, H.; Verboom, G. The importance of nutritional regulation of plant water flux. *Oecologia* **2009**, *161*, 15–24. [CrossRef]

57. Javaid, A. Arbuscular mycorrhizal mediated nutrition in plants. *J. Plant Nutr.* **2009**, *32*, 1595–1916. [CrossRef]

 *horticulturae*

Article

Trichoderma spp. and Mulching Films Differentially Boost Qualitative and Quantitative Aspects of Greenhouse Lettuce under Diverse N Conditions

Ida Di Mola [1,*], Lucia Ottaiano [1], Eugenio Cozzolino [2], Mauro Senatore [3], Adriana Sacco [4], Christophe El-Nakhel [1], Youssef Rouphael [1] and Mauro Mori [1]

[1]   Department of Agricultural Science, University of Naples Federico II, 80055 Portici Naples, Italy; lucyottaiano@gmail.com (L.O.); christophe.elnakhel@unina.it (C.E.-N.); youssef.rouphael@unina.it (Y.R.); mori@unina.it (M.M.)

[2]   Council for Agricultural Research and Economics (CREA)-Research Center for Cereal and Industrial Crops, 81100 Caserta, Italy; eugenio.cozzolino@crea.gov.it

[3]   Samagri Start up Innovativa srl, 84013 Cava de' Tirreni, Italy; mauro.senatore@samagri.it

[4]   National Council of Research-IPSP, 80055 Portici Naples, Italy; adriana.sacco@unina.it

*   Correspondence: ida.dimola@unina.it

Received: 31 July 2020; Accepted: 2 September 2020; Published: 4 September 2020

Abstract: The global increasing demand of lettuce is pushing farmers to boost their production through several technical means, including mulching and nitrogen fertilization. However, from an environmental protection perspective, the role of scientific research is to limit the excessive use of some chemical approaches. This research aims to evaluate the possible effects of two mulching films (black polyethylene, PE, and brown photoselective film, BF) and two treatments with a plant growth-promoting product, containing *Trichoderma* spp., (non-treated, - Control and treated with RYZO PEP UP, - TR), on the productive and qualitative traits of lettuce grown under four regimes of nitrogen (0, 30, 60 and 90 kg ha^{-1}, N0, N30, N60, and N90, respectively). The marketable yield increased at higher nitrogen levels, but without differences between the N60 and N90 doses. The photoselective film elicited marketable yield, with an 8% increase over PE. N fertilization also improved photochemical efficiency (higher Soil Plant Analysis Development and chlorophyllous pigments biosynthesis), as well as antioxidant activities (lipophilic—LAA and hydrophilic—HAA) and bioactive compounds (phenols and total ascorbic acid—TAA). Interestingly, *Trichoderma* spp. had a positive effect on these qualitative parameters, especially when combined with mulching films, where the increase generated by PE-TR treatment over the all other treatments was 16.3% and 16.8% for LAA and HHA, respectively. In all treatments, the nitrate leaves content was consistently always within the legal limit imposed by the European community. Overall, although *Trichoderma* spp. did not engender a marked effect on yield, probably due to the short crop cycle, its positive effect on some quality traits is an interesting starting point for further research.

Keywords: mulching film; *Trichoderma*; *Lactuca sativa* L.; nitrogen dose; nutritional quality; yield; sustainability

---

1. Introduction

Lettuce is the first fresh vegetable cultivated and commercialized in the world, especially in temperate and subtropical regions [1]. The increasing consumption of lettuce is driving farmers to intensify the use of chemical means in agricultural production, therefore not consistently achieving sustainable agriculture principles. In recent years, one of the used methods for boosting food production is plastic mulching films, which allow reaching higher yield in respect to bare soil, both under greenhouse or open-air conditions [2]. In fact, applying mulching films has pointed out several positive factors such as increase of soil temperature, moisture conservation, weeds and pests reduction, higher crop yields, and more efficient use of soil nutrients [3,4]. In agriculture, black polyethylene plastic films are normally used, but the photoselective plastic films have gained important consideration. In several studies, these films have showed very convincing results and performance in terms of quality and quantity of crop yields [5–7]. The traditional black PE film assure the mulching effect, blocking the solar radiation including the Photosynthetically Active Radiation (PAR). Instead, the photoselective films block only the PAR, hence exploiting the useful part of solar radiation to heat or "cool" the soil, based on the optical properties of each film [8].

In order to reach higher production, many crops including leafy vegetables require large quantities of nitrogen, even though its high nitrogen availability does not always correlate with a higher quality of the product [9]. Indeed, an overuse of this macronutrient can induce nitrate accumulation in leaves [10–12] with possible deleterious effects on human health [13]. Moreover, an excessive use of nitrogen can have a detrimental impact on the environment due to nitrate leaching into the groundwater and to greenhouse gas emissions [14,15]. Therefore, in order to overcome these adversities, it is necessary to adopt an adequate management of nitrogen fertilization through an equitable application of the requested dose, as well as to choose the convenient chemical form and application time [16].

Recently, many studies have proposed an eco-friendly approach to improve crop yield, which is the integrated use of plant biostimulants (PBs), especially the use of fungi, such as *Trichoderma*. These plant growth-promoting microorganisms improve pathogen/pest control, increase nutrient uptake, and stimulate photosynthesis and carbohydrate metabolism processes, consequently influencing positively crop productivity and quality [17,18].

Some authors [19–23] found that *Trichoderma* spp. acts as plant biostimulants, improving nutrient uptake, plant growth, and plant tolerance to abiotic stress. Moreover, Colla et al. [23], Fiorentino et al. [22], and Rouphael et al. [21] also demonstrated that the use of *Trichoderma* enhances nutrient-use efficiency (NUE) in lettuce, favoring the N uptake especially under sub-optimal nitrogen conditions.

Based on the abovementioned, several studies have been conducted regarding these factors, but their interaction is still not investigated. Thus, the aim of this research was to evaluate the possible effects of different mulching films and the application of *Trichoderma* as plant growth-promoting fungi on productive and qualitative traits of lettuce grown under different nitrogen regimes.

2. Materials and Methods

2.1. Experimental Setting, Design, and Plant Material

The trial was carried out during winter 2019–2020 in the soil of a polyethylene greenhouse at the experimental site "Gussone Park" of Department of Agricultural Science-DIA in Portici, Southern Italy. The tested crop was lettuce (*Lactuca sativa* L. var. *capitata*) cv. Jumper (Gautier Seed, Cesena, Italy), a plant with a semi-closed head and medium green brilliant leaves.

The design was a factorial comparison between (1) two mulching films, a brown photoselective film (DF) and a black polyethylene (PE); (2) two plant growth promoting treatments (non-treated, —Control and treated with RYZO PEP UP, - TR); (3) four increasing levels of nitrogen fertilization (0, 30, 60, and 90 kg ha^{-1}, N0, N30, N60, and N90, respectively). The optimal nitrogen dose was calculated by the balance method and it resulted 60 kg ha^{-1}. This method takes into account (i) the inputs, such as chemical soil fertility, available nitrogen from the mineralization of organic matter, rain, nitrates in irrigation water, and nitrogen released by previous crops, and (ii) the outputs, mainly crop uptakes, but also the leaching and the volatilization; therefore their difference gives the value of nitrogen needed.

The experimental design was a split plot design with three replications; experimental units were 48 and each one was 6 m long and 0.4 m large. The experimental soil was sandy loam, with neutral pH, a total nitrogen content of 1.2 g kg^{-1}, high content of potassium (1800 mg kg^{-1}) and phosphorus (85 mg kg^{-1}), and a good content of organic matter (1.7%).

2.2. Plant Management, Nitrogen Fertilization, and Trichoderma Application

In mid-January 2020, the mulching films were placed manually; then, on 21 January, the plants were immersed in a water solution with RYZO PEP UP at 1 mL L^{-1}, and the following day, the plants were transplanted at a density of 16 plants per square meter. RYZO PEP UP is a microbial biostimulant produced by Samagri SRL (Cava de' Tirreni, Salerno, Italy) containing *Trichoderma* spp. with a final concentration of 1×10^8 CFU. Moreover, lettuce plants were sprayed directly on the soil surroundings with a solution containing 3 mL L^{-1} of RYZO PEP UP, two times on a two-week basis starting 15 days after the sowing.

According to the experimental design, nitrogen was added as ammonium nitrate (26%) in a single operation two weeks after the transplant at the soil surface, respecting the dose calculated by the balance method. The water losses were calculated by the Hargreaves formula and were fully restored by irrigation.

2.3. Plant Growth Parameters, Marketable Yield, SPAD index, Leaf Colorimetry, and Nitrate Determination

Harvesting was done on three different dates (10, 12, and 17 March), when lettuce heads reached the commercial size. Ten plants per experimental plot (replicate) were harvested, the head diameter was measured, and the leaves were counted; all data were averaged and expressed as the mean of the three replicates. The marketable yield was expressed in tons per hectare. Before harvesting, the measurements of Soil Plant Analysis Development (SPAD) index were made by a portable chlorophyll meter SPAD-502 (Konica Minolta, Tokyo, Japan). On five fully expanded young leaves of five heads per each replicate, the color space parameters were measured by a Minolta CR-300 Chroma Meter (Minolta Camera Co. Ltd., Osaka, Japan) according to the Commission international de l'eclairage (CIELAB).

A representative fresh sample per replicate was stored at −80 °C for qualitative analysis. Meanwhile, another sample of each replicate was dried in oven at 70 °C until reaching constant weight, in order to be used afterwards for measuring the sum of nitrate and nitrite of NH$_4$Cl-buffer extract, by Foss FIAstar 5000 continuous flow analyzer. The method is based on the reduction of nitrate to nitrite on a cadmium reductor. The concentration of nitrite is negligible in most cases in comparison with nitrate concentration.

2.4. Chlorophyllous Pigments, Carotenoids and Bioactive Molecules Analysis

Chlorophyllous pigments and carotenoids were measured on 1 g of fresh sample based on the method of Lichtenhaler and Wellburn [24]. The samples were extracted with ammionacal acetone; then, through a spectrophotometer (Hach DR 2000, Hach Co., Loveland, CO, USA), the solution absorbances of chlorophyll a, chlorophyll b, and carotenoids were measured at 662, 647, and 470 nm, respectively, where total cholorphyll is the sum of chlorophyll a and b. The values were expressed as mg g^{-1} fresh weight (fw). The method of Kampfenkel et al. [25] was used for measuring total ascorbic acid content (TAA), which was expressed as mg ascorbic acid 100 g^{-1} fw. Meanwhile, the total phenolic content was determined by the Singleton et al. procedure [26] and expressed as mg gallic acid per 100 g^{-1} dry weight (dw).

2.5. Antioxidant Capacity Analysis

On 200 mg extract of freeze-dried leaves prepared through a freeze drier (Christ, Alpha 1-4, Osterode, Germany), the hydrophilic (HAA) and lipophilic (LAA) antioxidant activity were assessed by the N, N-dimethyl-p-phenylenediamine (DMPD) method [27] and ABTS (2,20-azinobis 3-ethylbenzothiazoline-6-sulfonic acid) method [28], respectively. The values were expressed as mmol ascorbic acid 100 g^{-1} dw for HAA and mmol of Trolox 100 g^{-1} dw for LAA.

2.6. Statistical Analysis

All data were analyzed by the SPSS software package (SPSS version 22, Chicago, IL, USA), using a general linear model (three-way ANOVA). Means were separated according to the Duncan Multiple Range Test (DMRT; significance level 0.05).

3. Results and Discussion

3.1. Effect of N Fertilization Dose, Trichoderma Application, and Mulching on Marketable Yield and Growth Parameters

Marketable yield was significantly affected only by nitrogen fertilization and mulching films; the interaction between the three experimental factors (mulching film; *Trichoderma* spp., nitrogen fertilization) was not detected. The nitrogen dose boosted the marketable yield that increased at higher nitrogen levels but without significant differences between the N60 and N90 treatments (Figure 1). Among the two mulching films, the photoselective film elicited marketable yield, with a 7.9% increase over PE.

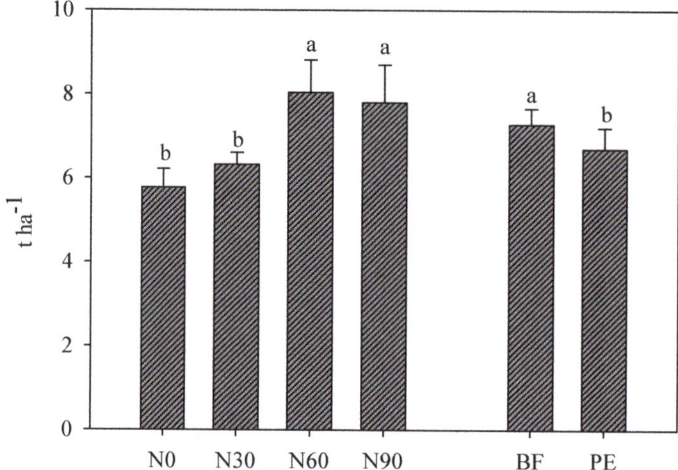

Figure 1. Lettuce yield in relation to mulching cover films (brown photoselective film—BF and black polyethylene—PE) and N fertilization rates (N0 = 0, N30 = 30, N60 = 60, and N90 = 90 kg ha^{-1}). Yield is presented as mean values of three replicates with relative standard errors. Different letters indicate significant differences at $p \leq 0.05$. All data are expressed as mean; n = 3.

In a perspective of sustainable agriculture, the scientific community is evaluating the possibility of using eco-friendly products, such as microbial biostimulants, including *Trichoderma* based product, to reduce undesirable footprint on the environment, due to an excessive or incorrect use of chemical means. In this study, *Trichoderma* did not have a positive effect on marketable yield, contrarily to the results of Fiorentino et al. [22] and Caruso et al. [20]. However, Fiorentino et al. [22] also found that there was no effect on rocket production, when treated by different *Trichoderma*-based biostimulants These results demonstrate that the response to *Trichoderma*-based products could be species-specific, particularly related to the different duration of crop cycles. In fact, we suppose that the effect of RYZO PEP UP on marketable yield was not significant, which was probably due to the lack of sufficient time for the fungi to establish in the soil/roots, boost the colonies and thus their activity. In effect, Chang et al. [29] also found a positive correlation between the colonies of *Trichoderma* in the soil and dry matter content in cucumber; therefore, we suppose that *Trichoderma* might need more time for fully colonizing the soil/roots zone. Instead, the recorded yield increase in lettuce grown on brown film could be due to specific thermal properties of this film. In fact, Guerrini et al. [30] found that the brown mulching film has very good thermal properties, which can guarantee higher temperatures (about +2–3 °C) than black film at the level of root, particularly at 5 to 10 cm depth, where the root system of lettuce is concentrated. This technical characteristic offers many agronomic benefits, especially for winter transplants, since the environmental conditions for plants are better; moreover, it limits the stress due to low temperatures and helps the plant to better form its root system. In a recent study, Bonanomi et al. [31] compared the effect of photoselective mulching film (yellow) to conventional black film, in order to verify their effects, both alone and combined with microbial consortia, on the yield of several crops including winter lettuce, where they found that the yellow photoselective film increased crop yield.

Regarding growth parameters (fresh weight, diameter, and leaf number), only the main effect of nitrogen fertilization and mulching film was found (Table 1). The photoselective mulching film improved all growth parameters, with 7.9%, 5.2%, and 8.8% for head fresh weight and diameter, and head leaf number, respectively. These growth parameters also increased gradually when N fertilization levels increased from 0 to 60 kg ha^{-1}, which showed +28.3%, +10.3%, and +15.0% over no fertilized plants (N0) for fresh weight, head diameter, and leaf number, respectively; instead, there were no significant differences among the N60 and N90 treatments (Table 1).

Table 1. Lettuce growth parameters in relation to mulching cover films (brown photoselective film —YF and a black polyethylene—PE), *Trichoderma* application (untreated—Control and treated—TR) and N fertilization rates (N0 = 0, N30 = 30, N60 = 60, and N90 = 90 kg ha^{-1}).

Treatments	Fresh Weight (g Plant^{-1})	Head Diameter (cm)	Number of Leaves (No. Plant^{-1})
Mulching			
BF	290.83 a	25.22 a	47.75 a
PE	267.90 b	23.89 b	43.54 b
Fertilization			
N0	230.80 c	23.15 b	41.50 c
N30	252.79 b	23.86 b	44.42 b
N60	321.74 a	25.82 a	48.83 a
N90	312.12 a	25.39 a	47.83 a
Trichoderma			
Control	281.65	24.30	46.04
TR	277.07	24.81	45.25
Significance			
Mulching (M)	**	*	*
Fertilization (F)	**	**	*
Trichoderma (TR)	NS	NS	NS
M × F	NS	NS	NS
M × TR	NS	NS	NS
F × TR	NS	NS	NS
M × F × TR	NS	NS	NS

NS, *, ** Non-significant or significant at $p \leq 0.05$ and 0.01. Different letters within each column indicate significant differences according to Duncan's test ($p \leq 0.05$). All data are expressed as mean; n = 3.

3.2. Effect of N Fertilization Dose, Trichoderma Application, and Mulching Films on Leaf Colorimetry and SPAD Index

The SPAD index is a key indicator of the nutritional status of plants, and its measurement is rapid and non-destructive. In this research, it was significantly affected by mulching films and nitrogen fertilization (Table 2). Particularly, the brown film slightly elicited the SPAD index compared to polyethylene film. Moreover, the SPAD index gradually increased when the nitrogen dose increased, but without differences between N60 and N90 treatments (Table 2): the average increase of fertilized over non-fertilized plants was 6.3%.

Table 2. Lettuce Soil Plant Analysis Development (SPAD) index and CIELAB color parameters (L *, a *, and b *) in relation to mulching cover films (brown photoselective film—YF and black polyethylene—PE), *Trichoderma* application (untreated—Control and treated—TR) and N fertilization rates (N0 = 0, N30 = 30, N60 = 60, and N90 = 90 kg ha^{-1}). [L* (lightness, ranging from 0 = black to 100 = white), a * [chroma component ranging from green (−60) to red (+60)], b * [chroma component ranging from blue (−60) to yellow (+60)].

Treatments	SPAD	L *	a *	b *
Mulching				
BF	38.62 a	49.89	−11.11	34.56
PE	37.79 b	48.95	−11.66	36.07
Fertilization				
N0	36.36 c	48.02 c	−10.69 a	36.84
N30	38.04 b	49.15 bc	−11.42 ab	35.73
N60	39.50 a	50.68 a	−11.61 bc	34.88
N90	38.93 a	49.83 ab	−11.82 c	33.81
Trichoderma				
Control	38.04	49.55	−11.44	34.90
TR	38.38	49.28	−11.31	35.73
Significance				
Mulching (M)	**	NS	NS	NS
Fertilization (F)	**	*	*	NS
Trichoderma (TR)	NS	NS	NS	NS
M × F	NS	NS	NS	NS
M × TR	NS	NS	NS	NS
F × TR	NS	NS	NS	NS
M × F × TR	NS	NS	NS	NS

NS, *, ** Non-significant or significant at $p \leq 0.05$ and 0.01. Different letters within each column indicate significant differences according to Duncan's test ($p \leq 0.05$). All data are expressed as mean; n = 3.

The product color is probably the main physical property that consumers observe when deciding a purchase. In the present study, the CIELAB parameters was only affected by nitrogen fertilization (Table 2). Interestingly, augmenting nitrogen fertilization from 0 to 90 kg ha^{-1}, the a* value significantly decreased, reflecting the increase of the green intensity, contemporary with the increase of brightness (L *; Table 2).

3.3. Effect of N Fertilization Dose, Trichoderma Application, and Mulching Films on Antioxidant Capacity and Bioactive Content

LAA, HAA, total phenols, and TAA were affected by nitrate fertilization; mulching films affected TAA, and finally, *Trichoderma* application affected HAA and TAA (Table 3). The interaction between the factors was never significant, except for the interaction of mulching films × *Trichoderma* regarding LAA (Figure 2) and HAA (Figure 3).

The antioxidant activity is one of the most important aspects in determining the nutritional quality of many foods, green leafy vegetables among them. The antioxidant molecules, such as ascorbic acid and phenols, have beneficial effects on human health, because they play a key role in delaying oxidative damage; therefore, they prevent several diseases [32,33].

The photoselective film enhanced total ascorbic acid; instead, there were no detected differences for the other parameters. LAA and HAA antioxidant activities, total phenols, and TAA showed a decreasing trend when nitrogen fertilization doses increased (Table 3). These results are in agreement with the results of Fiorentino et al. [22], who found a decrease in TAA values in lettuce leaves when N doses increased, and they are also in line with the findings of Di Mola et. [34], who found that HAA and TAA in baby lettuce decreased at high nitrogen application. Finally, Wang et al. [35] also observed a reduction of fruit and leafy vegetables quality (soluble solids and ascorbic acid) at high nitrogen fertilization doses. Probably, this behavior is due to the fact that the plants produce antioxidant compounds in response to stress; therefore, when nutritional condition is optimal, the plants show a lower antioxidant activity. Interestingly, *Trichoderma* improved HAA and TAA with an increase of 11.2% and 5.9%, respectively (Table 3). These findings are consistent with the results of Lombardi et al. [36], who applied three different *Trichoderma* bioactive metabolites on strawberry and found that one of them (HYTLO1) promoted the accumulation of ascorbic acid. Similarly, Rouphael et al. [21] and Caruso et al. [20] demonstrated an increase in TAA when *Trichoderma* was applied to lettuce and rocket, respectively. Nonetheless, HAA was as well higher in rocket treated with *Trichoderma* [20].

Table 3. Lipophilic (LAA) and hydrophilic (HAA) antioxidant activities, total phenols, and total ascorbic acid (TAA) in relation to mulching cover films (brown photoselective film–BF and black polyethylene–PE), Trichoderma application (untreated—Control and treated—TR) and N fertilization rates (N0 = 0, N30 = 30, N60 = 60, and N90 = 90 kg ha^{-1}).

Treatments	LAA (mM Trolox 100g^{-1} dw)	HAA (mM AA 100g^{-1} dw)	Phenols (mg gallic acid g^{-1} dw)	TAA (mg g^{-1} fw)
Mulching				
BF	9.52	6.69	1.552	19.59 a
PE	10.42	7.16	1.674	18.16 b
Fertilization				
N0	11.75 a	7.68 a	1.723 a	21.98 a
N30	10.70 ab	7.62 a	1.704 ab	20.76 b
N60	8.97 ab	6.14 b	1.576 bc	18.30 c
N90	8.48 b	6.24 b	1.448 c	14.47 d
Trichoderma				
Control	9.70	6.51 b	1.676	18.30 b
TR	10.24	7.33 a	1.549	19.45 a
Significance				
Mulching (M)	NS	NS	NS	*
Fertilization (F)	*	**	*	**
Trichoderma (T)	NS	*	NS	*
M × F	NS	NS	NS	NS
M × T	*	*	NS	NS
F × T	NS	NS	NS	NS
M × F × T	NS	NS	NS	NS

NS, *, ** Non-significant or significant at $p \leq 0.05$ and 0.01. Different letters within each column indicate significant differences according to Duncan's test ($p \leq 0.05$). All data are expressed as mean; n = 3.

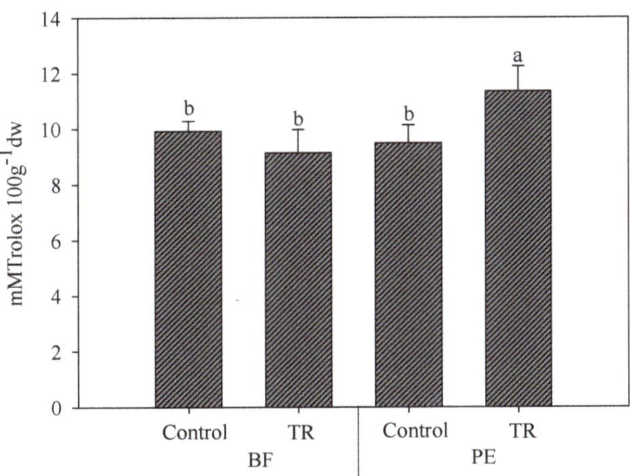

Figure 2. Effect of mulching films (brown photoselective film—BF and black polyethylene—PE) and *Trichoderma* application (untreated—Control and treated—TR) on lipophilic (LAA) antioxidant activity of lettuce leaves. All data are expressed as mean; n = 3.

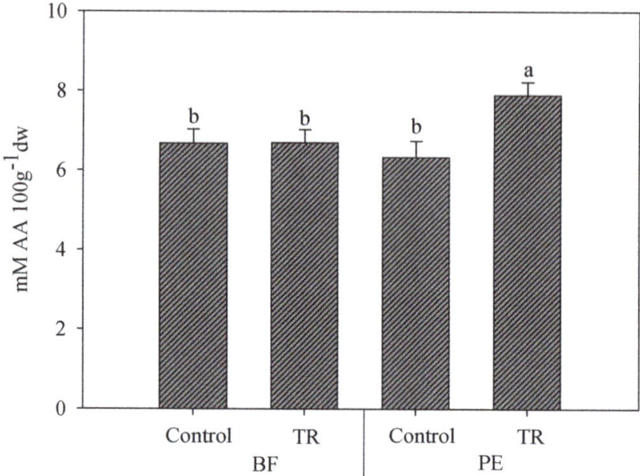

Figure 3. Effect of mulching films (brown photoselective film (BF) and black polyethylene (PE)) and *Trichoderma* application (untreated—Control and treated—TR) on hydrophilic (HAA) antioxidant activity of lettuce leaves. All data are expressed as mean; n = 3.

Regarding lipophilic and hydrophilic antioxidant activities, the combined effects of mulching film and application of *Trichoderma* were detected (Figures 2 and 3, respectively). Interestingly, for both qualitative parameters, the plants treated with *Trichoderma* and grown on black polyethylene showed the best performance, while the other three treatments were not different among them. In particular, the increase was 16.3% and 16.8% for LAA and HAA, respectively. Several studies report that *Trichoderma* grows best in a temperature range of 25 to 30 °C [37,38], and Guerrini et al. [30] found that black film reaches higher temperature in the soil–film gap, compared to other films (brown, transparent, and yellow film), since the surface of the black film heats up more than other films, and the heat is transferred directly to the contact layer. Therefore, since we sprayed *Trichoderma* solution exactly on the soil surface, we suppose that in this zone, the temperature conditions were better for the *Trichoderma* development and activity, and this probably boosted the antioxidant activity of lettuce.

3.4. Effect of N Fertilization Dose, Trichoderma Application, and Mulching Films on Biochemical Parameters and Nitrate Content

All parameters (chlorophyll a, b, total carotenoids and nitrate content) were affected by nitrogen fertilization doses. Moreover, chlorophyll b and nitrate content were also affected by *Trichoderma* application and total chlorophyll by mulching films (Table 4). Particularly, black PE enhanced chlorophyll (a, b and total), carotenoids, and nitrate content, but it was significant only for total chlorophyll, with an increase of 7.7% over photoselective film. Nitrogen fertilization positively affected the chlorophyll and carotenoids content but without significant differences among the nitrogen treatments. Chlorophyll a, b, total chlorophyll, and carotenoids increased on average +19.5%, 24.6%, 21.1%, and 15.8% compared to unfertilized plants, respectively (Table 4). Di Mola et al. [2,9] found a similar trend in baby lettuce and baby rocket regarding these parameters. However, the nitrogen fertilization negatively affected nitrate content in leaves, which increased from 0 to 60 kg N ha^{-1}, but without differences between N60 and N90. The increase in nitrate content is consistent with previous findings on baby leaf lettuce, baby rocket, lamb's lettuce, and spinach [9,16,34]. Finally, the effect of *Trichoderma* was evident on chlorophyll b (+12.8% over the control) and on leaves nitrate content, which was 8% more than control plants; however, this value was under the limit imposed by the European Union (EU) for lettuce marketing (Commission Regulation No. 1258/2011; 3000 to 5000 mg NO$_3^-$ kg^{-1}

of lettuce depending on growing season and cultivation conditions) [39]. The highest value of nitrate content is probably due to an easy uptake determined by *Trichoderma*, which also is able to solubilize Fe_2O_3, CuO, and metallic Zn, via (1) acidification by organic acids, (2) chelation by siderophores, (3) redox by ferric reductase, and (4) hydrolysis by phytase [40].

Table 4. Chlorophyll (a, b, and total), carotenoids, and nitrate content in relation to mulching cover films (brown photoselective film—YF and black polyethylene—PE), *Trichoderma* application (untreated—Control and treated—TR) and N fertilization rates (N0 = 0, N30 = 30, N60 = 60, and N90 = 90 kg ha^{-1}).

Treatments	Chlorophyll a (mg g^{-1} fw)	Chlorophyll b (mg g^{-1} fw)	Total Chlorophyll (mg g^{-1} fw)	Carotenoids (µg g^{-1} fw)	Nitrate (mg g^{-1} fw)
Mulching					
BF	0.552	0.224	0.776 b	293	1570.7
PE	0.599	0.242	0.841 a	309	1526.7
Fertilization					
N0	0.487 b	0.187 b	0.674 b	264 b	1163.8 c
N30	0.591 a	0.246 a	0.838 a	301 ab	1479.9 b
N60	0.606 a	0.246 a	0.852 a	316 ab	1794.3 a
N90	0.618 a	0.252 a	0.871 a	324 a	1756.7 a
Trichoderma					
Control	0.581	0.217 b	0.798	311	1483.9 b
TR	0.570	0.249 a	0.819	292	1613.5 a
Significance					
Mulching	NS	NS	*	NS	NS
Fertilization	*	*	*	*	**
Trichoderma	NS	*	NS	NS	*
M × F	NS	NS	NS	NS	NS
M × T	NS	NS	NS	NS	NS
F × T	NS	NS	NS	NS	NS
M × F × T	NS	NS	NS	NS	NS

NS, *, ** Non-significant or significant at $p \leq 0.05$ and 0.01. Different letters within each column indicate significant differences according to Duncan's test ($p \leq 0.05$). All data are expressed as mean; n = 3.

4. Conclusions

The photoselective film improved the marketable yield and growth parameters of lettuce (head weight, diameter, and leaf number), as well as the nitrogen fertilization (30 and 60 kg ha^{-1}). Moreover, nitrogen also enhanced the SPAD index, color parameters (brightness and green intensity), chlorophyll, and carotenoids content. The increasing levels of nitrogen also determined an increase in nitrate content in leaves but without overcoming the limit imposed by the European community. The application of *Trichoderma* had no positive effects on marketable yield and growth parameters, but it had an encouraging effect on antioxidant activity, especially when combined with polyethylene mulching film. Although the *Trichoderma* had no marked effect on yield, its positive effect on quality traits is an interesting starting point for further research. Its capacity to elicit a higher N uptake could improve production on long cycle-crops, simultaneously allowing a reduction of nitrogen application and mitigating environmental impact.

Author Contributions: Conceptualization, I.D.M. and M.M.; methodology, E.C.; software, I.D.M. and L.O.; validation, I.D.M., M.M., L.O., E.C., A.S., and M.S.; formal analysis, I.D.M. and L.O.; investigation, I.D.M. and L.O.; resources, I.D.M.; data curation, I.D.M., M.M., L.O., E.C., M.S., and A.S.; writing—original draft preparation, I.D.M.; writing—review and editing, I.D.M., C.E.-N. and Y.R.; visualization, M.M.; supervision, M.M.; project administration, I.D.M. and M.M.; funding acquisition, I.D.M. and M.M. All authors have read and agreed to the published version of the manuscript.

Funding: This research received no external funding.

Acknowledgments: We express our thanks to Roberto Maiello and Sabrina Nocerino for their technical assistance in the laboratory.

Conflicts of Interest: The authors declare no conflict of interest.

Abbreviations

SPAD	Soil Plant Analysis Development
LAA	Lipophilic Antioxidant Activity
HAA	Hydrophilic Antioxidant Activity
TAA	Total Ascorbic Acid

References

1. FAO. *The Future of Food and Agriculture–Alternative Pathways to 2050*; Food and Agriculture Organization of the United Nations: Rome, Italy, 2018.
2. Di Mola, I.; Cozzolino, E.; Ottaiano, L.; Duri, L.G.; Riccardi, R.; Spigno, P.; Mori, M. The effect of novel biodegradable films on agronomic performance of zucchini squash grown under open-field and greenhouse conditions. *Aust. J. Crop. Sci.* **2019**, *13*, 1810.
3. Kasirajan, S.; Ngouajio, M. Polyethylene and biodegradable mulches for agricultural applications: A review. *Agron. Sustain. Dev.* **2012**, *32*, 501–529. [CrossRef]
4. Kyrikou, I.; Briassoulis, D. Biodegradation of agricultural plastic films: A critical review. *J. Polym. Env.* **2007**, *15*, 125–150. [CrossRef]
5. Mormile, P.; Rippa, M.; Petti, L.; Immirzi, M.; Malinconico, M.B.; Lahoz, E.; Morra, L. Improvement of soil solarization through a hybrid system simulating a solar hot water panel. *J. Adv. Agric. Technol.* **2016**, *3*, 226–230. [CrossRef]
6. Mormile, P.; Capasso, R.; Rippa, M.; Petti, L. Light filtering by innovative plastic films for mulching and soil solarization: State of the art. *Acta Horticulturae* **2013**, *1015*, 113–121. [CrossRef]
7. Schettini, E.; De Salvador, F.R.; Scarascia Mugnozza, G.; Vox, G. Radiometric properties of photoselective and photoluminescent greenhouse plastic films and their effects on peach and cherry tree growth. *J. Hortic. Sci. Biothec.* **2011**, *86*, 79–83. [CrossRef]
8. Rippa, M.; Yan, C.; Petti, L.; Mormile, P. A simple method for saving water in agriculture based on the use of photo-selective mulch film. Available online: https://ijaer.in/2018files/ijaer_04__03.pdf (accessed on 3 September 2020).
9. Di Mola, I.; Ottaiano, L.; Cozzolino, E.; Senatore, M.; Giordano, M.; El-Nakhel, C.; Mori, M. Plant-based biostimulants influence the agronomical, physiological, and qualitative responses of baby rocket leaves under diverse nitrogen conditions. *Plants* **2019**, *8*, 522. [CrossRef]
10. Cavaiuolo, M.; Ferrante, A. Nitrates and glucosinolates as strong determinants of the nutritional quality in rocket leafy salads. *Nutrients* **2014**, *14*, 1519–1538. [CrossRef]
11. Alberici, A.; Quattrini, E.; Penati, M.; Martinetti, L.; Gallina, P.M.; Ferrante, A.; Schiavi, M. Effect of the reduction of nutrient solution concentration on leafy vegetables quality grown in floating system. *Acta Horticulturae* **2008**, *801*, 1167–1176. [CrossRef]
12. Wang, Z.; Li, S. Effects of nitrogen and phosphorus fertilization on plant growth and nitrate accumulation in vegetables. *J. Plant. Nutr.* **2004**, *27*, 539–556. [CrossRef]
13. Colla, G.; Kim, H.J.; Kyriacou, M.C.; Rouphael, Y. Nitrate in fruits and vegetables. *Sci. Hortic.* **2018**, *237*, 231–238. [CrossRef]
14. Lassaletta, L.; Billen, G.; Grizzetti, B.; Anglade, J.; Garnier, J. 50 year trends in nitrogen use efficiency of world cropping systems: The relationship between yield and nitrogen input to cropland. *Env. Res. Lett.* **2014**, *9*, 105011. [CrossRef]
15. Sutton, M.A.; Bleeker, A.; Howard, C.M.; Bekunda, M.; Grizzetti, B.; de Vries, W.; van Grinsven, H.J.M.; Abrol, Y.P.; Adhya, T.K.; Billen, G.; et al. Our Nutrient World: The challenge to produce more food and energy with less pollution. *Cent. Ecol. Hydrol.* **2013**, *8*, 95–108.
16. Di Mola, I.; Cozzolino, E.; Ottaiano, L.; Nocerino, S.; Rouphael, Y.; Colla, G.; Mori, M. Nitrogen Use and Uptake Efficiency and Crop Performance of Baby Spinach (*Spinacia oleracea* L.) and Lamb's Lettuce (*Valerianella locusta* L.) Grown under Variable Sub-Optimal N Regimes Combined with Plant-Based Biostimulant Application. *Agronomy* **2020**, *10*, 278. [CrossRef]
17. López-Bucio, J.; Pelagio-Flores, R.; Herrera-Estrella, A. Trichoderma as biostimulant: Exploiting the multilevelproperties of a plant beneficial fungus. *Sci. Hortic.* **2015**, *196*, 109–123.

18. Woo, S.; Ruocco, M.; Vinale, F.; Nigro, M.; Marra, R.; Lombardi, N.; Pascale, A.; Lanzuise, S.; Manganiello, G.; Lorito, M. Trichoderma-based products and their widespread use in agriculture. *Open Mycol. J.* **2014**, *8*. [CrossRef]

19. Carillo, P.; Woo, S.L.; Comite, E.; El-Nakhel, C.; Rouphael, Y.; Fusco, G.M.; Vinale, F. Application of *Trichoderma harzianum*, 6-pentyl-α-pyrone and plant biopolymer formulations modulate plant metabolism and fruit quality of plum tomatoes. *Plants* **2020**, *9*, 771. [CrossRef]

20. Caruso, G.; El-Nakhel, C.; Rouphael, Y.; Comite, E.; Lombardi, N.; Cuciniello, A.; Woo, S.L. *Diplotaxis tenuifolia* (L.) DC. Yield and Quality as Influenced by Cropping Season, Protein Hydrolysates, and Trichoderma Applications. *Plants* **2020**, *9*, 697. [CrossRef]

21. Rouphael, Y.; Carillo, P.; Colla, G.; Fiorentino, N.; Sabatino, L.; El-Nakhel, C.; Cozzolino, E. Appraisal of Combined Applications of *Trichoderma virens* and a Biopolymer-Based Biostimulant on Lettuce Agronomical, Physiological, and Qualitative Properties under Variable N Regimes. *Agronomy* **2020**, *10*, 196. [CrossRef]

22. Fiorentino, N.; Ventorino, V.; Woo, S.L.; Pepe, O.; De Rosa, A.; Gioia, L.; Romano, I.; Lombardi, N.; Napolitano, M.; Colla, G.; et al. Trichoderma-based biostimulants modulate rhizosphere microbial populations and improve N uptake efficiency, yield, and nutritional quality of leafy vegetables. *Front. Plant. Sci.* **2018**, *9*, 743. [CrossRef] [PubMed]

23. Colla, G.; Rouphael, Y.; Di Mattia, E.; El-Nakhel, C.; Cardarelli, M. Co-inoculation of Glomus intraradices and *Trichoderma atroviride* acts as a biostimulant to promote growth, yield and nutrient uptake of vegetable crops. *J. Sci. Food Agric.* **2015**, *95*, 1706–1715. [CrossRef] [PubMed]

24. Lichtenhaler, H.K.; Wellburn, A.R. Determinations of total carotenoids and chlorophylls a and b of leaf extracts in different solvents. *Biochem. Soc. Trans.* **1983**, *11*, 591–592. [CrossRef]

25. Kampfenkel, K.; Montagu, M.V.; Inzè, D. Extraction and determination of ascorbate and dehydroascorbate from plant tissue. *Annu. Rev. Biochem.* **1995**, *225*, 165–167. [CrossRef]

26. Singleton, V.L.; Orthofer, R.; Lamuela-Raventos, R.M. Analysis of total phenols and other oxidation substrates and antioxidants by means of Folin–Ciocalteu reagent. In *Methods in Enzymology*; Academic Press: Cambridge, MA, USA, 1999; Volume 299, pp. 152–178.

27. Fogliano, V.; Verde, V.; Randazzo, G.; Ritieni, A. Methods for measuring antioxidant activity and its application to monitoring the antioxidant capacity of wines. *J. Agric. Food Chem.* **1999**, *47*, 1035–1040. [CrossRef]

28. Pellegrini, N.; Re, R.; Yang, M.; Rice-Evans, C. Screening of dietary carotenoids and carotenoid-rich-fruit extracts for antioxidant activities applying 2,2- Azinobis (3-ethylenebenzothiazoline-6-sulfonic acid radical cation decoloration assay. In *Methods in Enzymology*; Academic Press: Cambridge, MA, USA, 1999; Volume 299, pp. 379–384.

29. Chang, Y.C.; Chang, Y.C.; Baker, R.; Kleifeld, O.; Chet, I. Increased growth of plants in the presence of the biological control agent *Trichoderma harzianum*. *Plant. Dis.* **1986**, *70*, 145–148. [CrossRef]

30. Guerrini, S.; Yan, C.; Malinconico, M.; Mormile, P. Agronomical overview of mulch film systems. In *Polymers for Agri-Food Applications*; Springer: Cham, Switzerland, 2019; pp. 241–264.

31. Bonanomi, G.; Chirico, G.B.; Palladino, M.; Gaglione, S.A.; Crispo, D.G.; Lazzaro, U.; Rippa, M. Combined application of photo-selective mulching films and beneficial microbes affects crop yield and irrigation water productivity in intensive farming systems. *Agric. Water Manag.* **2017**, *184*, 104–113. [CrossRef]

32. Kyriacou, M.C.; Rouphael, Y. Towards a new definition of quality for fresh fruits and vegetables. *Sci. Hortic.* **2018**, *234*, 463–469. [CrossRef]

33. Khanam, U.K.S.; Oba, S.; Yanase, E.; Murakami, Y. Phenolic acids, flavonoids and total antioxidant capacity of selected leafy vegetables. *J. Fun. Foods* **2012**, *4*, 979–987. [CrossRef]

34. Di Mola, I.; Cozzolino, E.; Ottaino, L.; Giordano, M.; Rouphael, Y.; Colla, G.; Mori, M. Effect of Vegetal- and Seaweed Extract-based Biostimulants on Agronomical and Leaf Quality Traits of Greenhouse Baby Lettuce under Four Regimes of Nitrogen Fertilization. *Agronomy* **2019**, *10*, 571. [CrossRef]

35. Wang, Z.H.; Li, S.X.; Malhi, S. Effects of fertilization and other agronomic measures on nutritional quality of crops. *J. Sci. Food Agric.* **2008**, *88*, 7–23. [CrossRef]

36. Lombardi, N.; Salzano, A.M.; Troise, A.D.; Scaloni, A.; Vitaglione, P.; Vinale, F.; Lanzuise, S. Effect of trichoderma bioactive metabolite treatments on the production, quality and protein profile of strawberry fruits. *J. Agric. Food Chem.* **2020**, *68*, 7246–7258. [CrossRef]

37. Petrisor, C.; Paica, A.; Constantinescu, F. Influence of abiotic factors on in vitro growth of Trichoderma strains. *Proc. Rom. Acad. Ser. B* **2016**, *18*, 11–14.

38. Singh, A.; Shahid, M.; Srivastava, M.; Pandey, S.; Sharma, A.; Kumar, V. Optimal physical parameters for growth of Trichoderma species at varying pH, temperature and agitation. *Virol. Mycol.* **2014**, *3*, 127–134.
39. European Community. Reg. n° 1258 of 2 December 2011. *Off. J. Eur. Union* **2011**, *L 320*, 15–17.
40. Li, R.X.; Cai, F.; Pang, G.; Shen, Q.R.; Li, R.; Chen, W. Solubilisation of phosphate and micronutrients by *Trichoderma harzianum* and its relationship with the promotion of tomato plant growth. *PLoS ONE* **2015**, *10*, e0130081. [CrossRef]

Article

Effects of Municipal Solid Waste Compost Supplemented with Inorganic Nitrogen on Physicochemical Soil Characteristics, Plant Growth, Nitrate Content, and Antioxidant Activity in Spinach

Rui M. A. Machado [1],*, Isabel Alves-Pereira [2],*, Miguel Robalo [1] and Rui Ferreira [2]

[1] MED—Mediterranean Institute for Agriculture, Environment and Development & Departamento de Fitotecnia, Escola de Ciências e Tecnologia, Universidade de Évora, 7002-554 Évora, Portugal; migueltbrobalo@gmail.com

[2] MED—Mediterranean Institute for Agriculture, Environment and Development & Departamento de Química, Escola de Ciências e Tecnologia, Universidade de Évora, 7002-554 Évora, Portugal; raf@uevora.pt

* Correspondence: rmam@uevora.pt (R.M.A.M.); iap@uevora.pt (I.A.-P.)

Citation: Machado, R.M.A.; Alves-Pereira, I.; Robalo, M.; Ferreira, R. Effects of Municipal Solid Waste Compost Supplemented with Inorganic Nitrogen on Physicochemical Soil Characteristics, Plant Growth, Nitrate Content, and Antioxidant Activity in Spinach. *Horticulturae* **2021**, *7*, 53. https://doi.org/10.3390/horticulturae7030053

Academic Editor: Nazim Gruda

Received: 24 February 2021
Accepted: 15 March 2021
Published: 17 March 2021

Publisher's Note: MDPI stays neutral with regard to jurisdictional claims in published maps and institutional affiliations.

Abstract: In this study, we evaluated the effects of municipal solid waste compost supplemented with inorganic N on the physicochemical properties of soil, plant growth, nitrate concentration, and antioxidant activity in spinach. Experiments were carried out in neutral and acidic soils that were low in organic matter. A fertilized soil was used as a control, while four compost treatments—two compost rates of 35 and 70 t ha^{-1}, supplemented or not with inorganic N (92 kg N ha^{-1} as Ca (NO$_3$)$_2$)—were applied by fertigation. The addition of compost increased the soil organic matter content and pH in both soils. The compost supplementation with N greatly increased the shoot dry weight and spinach fresh yield by nearly 109%. With the highest compost rate and 43% N applied, the yield increased in both soils, similar to results obtained in fertilized soil (3.8 kg m^{-2}). The combined application of compost and N could replace inorganic P and K fertilization to a significant extent. The compost application at both rates and in both soils considerably decreased shoot Mn concentrations.

Keywords: soil organic matter; acidic soil; pH; nitrogen; nutrient uptake; photosynthetic pigments; antioxidant activity; *Spinacia oleracea*

1. Introduction

Soil organic matter decline and soil acidity are global problems for crop production. Almost half of the European soils have low organic matter content, principally in southern Europe, France, the United Kingdom, and Germany [1,2]. The decline of soil organic matter can be accentuated by climate change and the increase of intensive farming. Soil acidification affects up to 40% of the world's arable soils [3,4].

In Portugal, most soils have low organic matter content due to climatic conditions, poor agricultural practices, and low soil pH [5]. In these soils, plants grow poorly because of low water availability. In addition, the combination of H$_3$O$^+$, Al, and Mn toxicities lead to a lack of essential nutrients [3,6]. Spinach plants grown in these soils without fertilization show reduced growth and leaf chlorosis, probably due to a lack of nutrients, especially nitrogen, resulting in plant death [7,8].

Municipal organic wastes, when collected separately and properly composted, produce high-quality municipal solid waste compost (MSWC) for agriculture, with low heavy metal content and high organic matter content [9–11].

The separate collection of bio-waste and compost is increasing in European Union countries. All EU Member States will be obliged to collect bio-waste separately in the coming years [12]. MSWC can be used to preserve and enhance SOM pools and reduce soil acidity and inorganic nutrient inputs. Soil pH is one the most decisive factors affecting plant nutrition, metal solubility, nutrient movement, and microbial activity. Compost application generally increases soil pH [13]. Increased soil pH is regarded as a major advantage when MSW compost is used [14].

MSWC can reduce soil acidity by increasing hydronium (H_3O^+) concentrations in soil, since mature MSW composts usually have high pH [10], with adsorption of organic anions and the corresponding release of hydroxyl ions [15]. Soil organic matter offers many negatively charged sites to bind H_3O^+ in acidic soil or from which to release H_3O^+ in basic soil, in both cases pushing soil solutions towards neutral [16]. The main constraint to plant growth in soils amended with MSW compost is soil nitrogen availability [7,14,17], because nitrogen is released from MSWC slowly and irregularly. In order to enhance N availability and avoid reductions in crop yield, the addition of compost must be supplemented by inorganic nitrogen.

Therefore, the aim of this study was to evaluate the effects of the rate of MSWC supplemented or not with inorganic nitrogen on the physicochemical properties of the two soils with low organic matter content (a neutral (pH 7.1) and an acidic (pH 5.5)) on plant growth, nitrate concentration, and antioxidant activity in spinach.

2. Materials and Methods

2.1. Growth Conditions

The study was conducted in a greenhouse located at the "Herdade Experimental da Mitra" (38°31'52" N; 8°01'05" W), University of Évora, Portugal. The greenhouse was covered with polycarbonate and had no supplemental lighting. Air temperatures inside the greenhouse ranged from 5 to 35 °C (Figure 1) and outside solar radiation ranged from 76.6 to 262.8 $W \cdot m^{-2} \cdot d^{-1}$ [18].

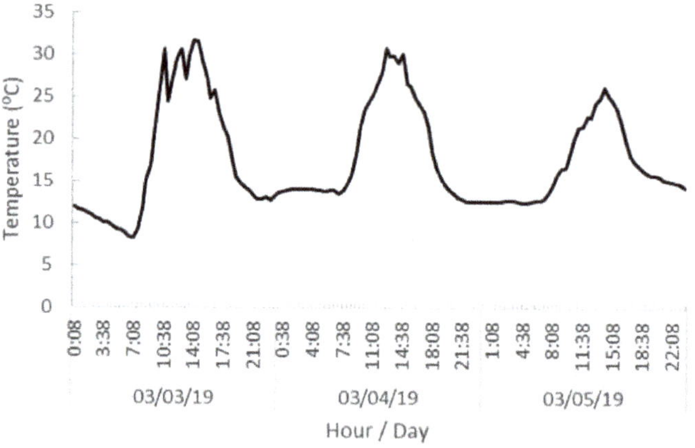

Figure 1. Diurnal changes in air temperature inside the greenhouse at the plant canopy level. The pattern illustrated is for temperatures measured from 3 to 5 March.

The experiment was carried out with two soils with low organic matter (a sand, loamy, neutral soil and a loamy sand soil that was strongly acid) (Table 1), five treatments (fertilized soil (FS) and four MSW compost treatments), two rates of MSW compost (35 and 70 t ha^{-1}), and the same rates of MSWC supplemented by nitrogen applied weekly in fertigation (35 + N, 70 + N). The rate of 30 t MSWC was chosen because it is a common rate of application. The highest rate (75 t MSWC) was calculated so as not to surpass the maximum amount of heavy metals that can be incorporated annually in the soils [19]. For the fertilized soil this was applied by fertigation 1.05 g N/pot (184 kg N ha^{-1}), while for the 35 + N and 70 + N half that amount was applied (0.53 g N/pot; 92 kg N ha^{-1}). Treatments were arranged in a randomized complete block design with six replicate pots per treatment.

In the experiments, we did not include the unfertilized treatment soil, since in previous experiments spinach plants in these unfertilized soils showed reduced growth and leaf chlorosis, probably because of lack of nutrients, especially nitrogen, resulting in plant death [7,8].

The experiment was carried out in plastic pots. Each 12 L plastic pot (21 cm height $\times$ 26.5 cm diameter) was filled with $\approx$14 kg of the soil from the upper layer (0–25 cm) of two different soil types obtained from the Mitra Research Farm in Évora, Portugal. The main characteristics of the soils are presented in Table 1. Ten days prior to transplanting, mature municipal solid waste organic compost (Nutrimais, Lipor, Lda, Portugal) in pellet form was added to each pot and mixed with the upper 10 cm of the soil.

In the fertilized soil (FS), we incorporated in the upward 10 cm of soil 0.17 g N, 0.35 g P_2O_5, 0.52 g K_2O, and 0.035 g MgO.

Table 1. The physicochemical properties of the soils.

	Soil	
	Neutral	**Acidic**
pH	7.19	5.50
Organic matter (%)	1.62	1.10
EC_e ($dS\ m^{-1}$)	0.082	0.03
Bulk density (g cm^{-3})	1.39	1.47
NO_3^- (ppm)	43.6	20.5
P_2O_5 (ppm)	238.0	10.0
K_2O (ppm)	204.0	60.0
Ca (meq $100\ g^{-1}$)	8.34	1.16
Mg (meq $100\ g^{-1}$)	1.20	0.27
Na (meq $100\ g^{-1}$)	0.13	0.70
CEC (meq $100\ g^{-1}$)	9.39	5.70
K (meq $100\ g^{-1}$)	0.49	0.11
Texture	Sand loamy	Loamy sand
Sand (%)	70.3	81.2
Loam (%)	12.3	8.00
Clay (%)	17.4	10.8

The raw materials used in the "Nutrimais" manufacturing process include horticultural products; food scraps carefully selected from restaurants, canteens, and similar establishments; forest exploitation residues (e.g., branches and foliage); and green residues (e.g., flowers, grasses, prunings). The physicochemical characteristics of the MSWC are presented in Table 2. The maximum heavy metal concentration of the MSWC was low. It was below the maximal values for the Portuguese legislation for class I compost [19] and for class A in different European countries [9].

Table 2. Physicochemical characteristics of the municipal solid waste compost (MSWC) and maximum values of heavy metals in class I organic composts.

Municipal Soil Waste Compost				Maximum Values of Heavy Metals for Organic Composts of Class I [3]	
pH [1]	8.68	Cd ($mg.kg^{-1}$)	0.35	Cd ($mg.kg^{-1}$) [3]	0.7
EC (Elec. Conductivity) ($dS.m^{-1}$) [1]	5.4	Pb ($mg.kg^{-1}$)	32	Pb ($mg.kg^{-1}$)	100
Organic Matter (%) [2]	52.47	Cr ($mg.kg^{-1}$)	22.3	Cr ($mg.kg^{-1}$)	100
Humidity (%)	10.52	Cu ($mg.kg^{-1}$)	49	Cu ($mg.kg^{-1}$)	100
Humic acids (%)	3.71	Hg ($mg.kg^{-1}$)	0.1	Hg ($mg.kg^{-1}$)	0.7
Ratio C/N	11.94	Ni ($mg.kg^{-1}$)	7.47	Ni ($mg.kg^{-1}$)	50
C (%)	29.15	Zn ($mg.kg^{-1}$)	160	Zn ($mg.kg^{-1}$)	200
N (%)	2.41	B ($mg.kg^{-1}$)	38	B($mg.kg^{-1}$)	
P_2O_5 (%)	1.49				
K_2O (%)	1.81				
CaO (%)	15.17				
MgO (%)	0.7				

[1] EC and pH were measured in extracted 1:5 compost/water, *w/v*. [2] Concentrations are expressed on a dry weight basis. The moisture of the compost before soil application was 14%. [3] Portuguese legislation [19].

Soil-blocked spinach (*Spinacia oleracea* L. cv. Manatee) seedlings (seven seedlings per block, three blocks per pot = 339 plants m^{-2}) were transplanted (19 February 2019) after 18 days following emergence into 12 L pots.

Plants were watered by hand daily (9–10 am) to avoid applying high volumes of water, minimizing drainage losses and preventing plants from suffering water stress.

The volume of water applied (ranged from 90 to 400 mL/pot) was adjusted to the climatic conditions (temperature and solar radiation), readings of the volumetric soil water content, and the soil water storage capacity of the soils. Volumetric soil water content was measured daily (08:00–09:00) using a soil moisture probe (SM105T delta devices UK). The irrigation water had a low EC_w (0.1 dS m^{-1}).

Nitrogen was applied via fertigation once a week in five equal fertilizer applications, starting at transplantation and finishing in the week before harvest. The fertilizer used to apply nitrogen was calcium nitrate (15.5% $N\text{-}NO_3$, 1.1% $N\text{-}NH_4$, and 26.5% CaO). The nutrients applied using inorganic fertilizers and MWSC in each treatment are presented in Table 3.

Table 3. Total amounts of nutrients added in each treatment.

Treat.	Type	N	P_2O_5	K_2O	CaO	MgO
				Kg ha^{-1}		
FS	Inorg. [1]	214.1 [3]	61.4	91.3	294 [4]	6.14
35	Org. [2]	679.7	431.6	524.3	4545.2	202.7
70	Org.	1395.4	862.7	1047.9	9090.3	405.3
35 + N	Org. + Inorg.	679.7 + 92	431.6	524.3	4545.2 + 147 [4]	202.7
70 + N	Org. + Inorg.	1395.4 + 92	862.7	1047.9	9090.3 + 147 [4]	405.3

[1] Inorganic fertilizer; [2] MSWC; [3] 29.8 kg N ha^{-1} applied before spinach plantation and 184 kg N ha^{-1} applied via fertigation; [4] applied in fertigation.

Starting from transplantation, air temperature at the plant canopy level was monitored hourly using a T-Logger HI141 temperature sensor (Hanna Instruments) (Figure 1). The weeds were regularly manually removed from the pots.

2.2. Measurements

The plants were harvested 38 days after transplantation. The shoots of the plants were cut off 1 cm above the substrate surface. Ten representative plants (shoots) from each box were washed, oven-dried at 70 °C for 2–3 days, weighed, ground so that they would pass through a 40-mesh sieve, then analyzed for N, P, K, Ca, Mg, Na, B, Cu, Mn, and Zn. Total N was analyzed by using a combustion analyzer (Leco Corp. St. Josef, MI, USA). The K and Na were analyzed by flame photometry (Jenway, Dunmow, UK). The P and B were analyzed using a UV/Vis spectrometer (Perkin Elmer lamba25). The remaining nutrients were analyzed using an atomic absorption spectrometer (Perkin Elmer, Inc., Shelton, CT, USA)

After harvesting of plants, three soil cores were collected at random from each pot using a soil probe measuring 3 cm in diameter and 0.1 m in depth in order to analyze soil, $NO_3\text{-}N$, pH, and electrical conductivity (EC_e) and organic matter content.

Soil nitrate was measured using an ion-specific electrode and meter (Crison Instruments, Barcelona, Spain), using the method outlined by [20]. Soil pH was measured in 1:2.5 soil/water suspensions using a potentiometer (pH Micro 2000 Crison). EC_e was measured in 1:5 soil/water aqueous extracts using a conductivity meter (LF 330 WTW, Weilhein, Germany). Organic carbon (%) was measured using a sulfur and carbon determinator (SC-144 DR, Leco Inc, St. Joseph, MI, USA). Organic matter (%) was estimated from organic carbon (%) using the conversion factor 1.72 (organic matter (%) = total organic carbon (%) × 1.72) [21].

Leaf samples from ten treatments and five replicates were stored at −80 °C for NO_3^- determination according to [22]. Briefly, portions (0.1000 g) of spinach leaves were suspended in 10 mL of distilled water. The samples were oven-dried at 65 °C for 48 h, macerated in a mortar, homogenized in a test tube with 10 mL of distilled water, agitated in a vortex, and incubated for 1 h at 45 °C in a shaking water bath. Filtrated extracts in Whatman 40 filter paper were then mixed with salicylic acid in 5% sulfuric acid (1:4), incubated for 20 min at room temperature, and mixed with 9.5 mL of 2 M sodium hydroxide. The concentration of NO_3^- in the solution was then determined by reading the absorbance at 338 and 440 nm using a calibration curve (NO_3^-, $n = 6$ concentrations between 0 and 500 mg/L).

In order to determine the photosynthetic pigment content, 1.000 g of spinach leaf from each treatment was macerated in a mortar and homogenized in 8 mL of methanol/water solution (90:10 (v/v), M90-extract) for 1 min, then centrifuged at 4 °C at $6440 \times g$ for 5 min. Chlorophyll a and b and carotenoids were quantified in aliquots of M90-extract via UV-Vis spectrophotometry, using the appropriate equations [23]:

Chl a (μg/mL) = 16.82 A665.2 − 9.28 A652.4;

Chl b (μg/mL) = 36.92 A652.4 − 16.54 A665.2;

Cc (μg/mL) = (1000 A470 − 1.91Chl a − 95.15Chl b)/225.

where A = Absorbance, Chl a = Chlorophyll a, Chl b = Chlorophyll b, Cc = carotenoids.

In order to determine free radical scavenging antioxidant activity (DPPH), 1000 g of leaf sample from each treatment was macerated in a mortar and homogenized in 8 mL of methanol/water solution (80:20, v/v) for 1 min, then centrifuged for 5 min at 4 °C and $6440 \times g$ (M80 extract). Aliquots of methanol extracts were stored at −20 °C for later use. Antioxidant activity was determined by measuring the ability of M80 spinach extracts to scavenge the violet-colored stable organic radical 2,2-diphenyl-1-picryl-hydrazyl (DPPH$^\bullet$), converting it into the yellow-colored stable product diphenyl-picryl hydrazine (DPPH-H). Aliquots of an extemporaneous methanol solution of 0.03 g/L DPPH$^\bullet$, which were kept in the dark, were added to a known volume of sample (M80 extract) or standard solution. The reduction of DPPH$^\bullet$ to DPPH-H was followed by reading the absorbance at 515 nm and 25 °C for 180 s. Antioxidant activity, reported as milligrams of GAE (gallic acid equivalent) per 100 g of FW, was calculated using a calibration curve (GAE, n = 8 concentrations from 0 to 200 mg·L^{-1}) [24].

2.3. Data Analysis

Data were processed via analysis of variance using SPSS Statistics 25 software (Chicago, IL, USA), licensed to the University of Évora. Means were separated at the 5% level using Duncan's new multiple range test.

3. Results and Discussion

3.1. Soil Physicochemical Properties

Soil moisture values at depths of 0 to 5 cm were not affected by the interactions between the treatments. The moisture content values for all samplings were significantly greater in the neutral soil (Figure 2). The addition of MSWC to the soil, as compared to fertilized soil, significantly increased the moisture contents at 24 and 31 days after plantation (DAP) (Figure 2). On the last sampling date, soil moisture content increased up to 12.3 %. Moisture content was not affected by the MSWC rate when combined or not with nitrogen. This could be due to a higher plant water uptake caused by the increase of the yield due to the combined application of MSWC and nitrogen (Table 4).

Soil temperature values at 10 cm depth were not significantly affected by the treatments, nor by their interaction. The temperature values measured at 10:00 to 11:30 on the different sampling dates ranged from 18.0 to 24.5 °C, probably increasing during the day. These temperatures are favorable for organic matter mineralization (Figure 2).

Soil organic matter content (SOM), pH, EC_e, and nitrate values were significantly affected by the interactions between treatments ($p < 0.001$), indicating that the responses of soil to the addition of MSW compost differed. Despite the SOM, pH, and EC_e in both soils, in relation to fertilized soil, these components increased with the addition of MSWC supplemented or not with inorganic nitrogen (Figure 3). In neutral soil, SOM increased with the rate of MSWC (Figure 3a). However, in acidic soil, SOM only increased significantly with MSWC when supplemented with inorganic nitrogen (Figure 3a).

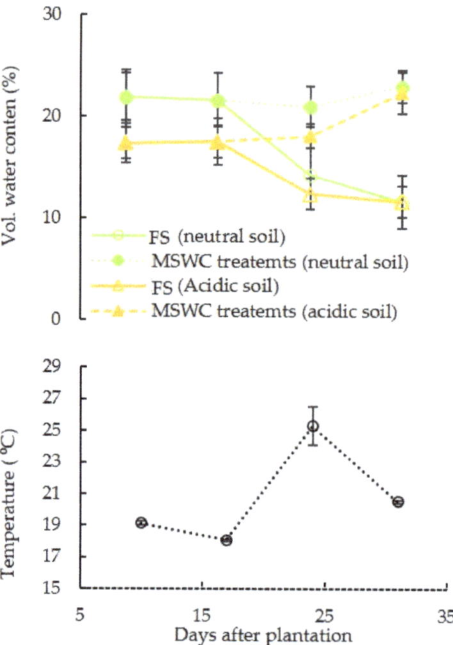

Figure 2. Effects of soil and MSWC supplemented or not with inorganic nitrogen on soil volumetric water content (%) values at 0–5 cm depth and soil temperature at 10 cm depth. Each symbol represents the mean of six replicates, while the error bars represent ±1 standard error.

The SOM values in neutral and acidic soils with the addition of 70 t MSWC ha^{-1} were higher than 3.5 and 2%, respectively. Regarding soil before spinach plantation, the addition of 35 and 70 t MSWC to neutral soil increased the average SOM contents by 1.15 and 2.25%, respectively. However, in acidic soil, for the same rates of MSWC the increases of SOM were only 0.45 and 0.90% respectively. This could be because of the soil characteristics (e.g., bulk density) or the different organic matter decomposition rates.

The addition of MSWC to the soils at both rates, supplemented or not with inorganic nitrogen, increased soil pH values relative to those before plantation and only fertilized with nitrogen (Figure 3b). Increases in soil pH with the addition of MSWC have also been reported by other authors [11,25].

However, regarding neutral soil, the amount of compost had no significant influence on soil pH, while in the acidic soil it significantly increased with the amount of the compost (Figure 3b). Paradelo and Barral (2017) also reported that the soil pH of acidic soils increased with the addition of MSWC [26].

Soil pH increases in the neutral and acidic soils ranged on average from 0.61 to 0.89 and from 1.09 to 1.85, respectively (Figure 3b). The difference in the magnitude of the pH increases may be due to the initial soil pH and cation exchange capacity (CEC) of the soils, and also due to the increase in the soil buffer capacity caused by the addition of MSWC. The humic acids in MSWC intensify the CEC and buffering capacity of the soil [27].

The addition of 35 and 70 t of MSWC ha^{-1} to neutral soil increased the soil pH from 7.17 to average values of 7.7 and 8.0, respectively (Figure 3b). These values can negatively affect plant nutrition, since they can decrease the nutrient availability in the soil solution and can reduce organic matter mineralization, since neutral or slightly alkaline conditions favor bacterial growth [28].

Therefore, regarding the neutral soil, the rates of MSWC used can be higher. Conversely, soil pH values in acidic soil increased from 5.5 to average values of 6.8 and 7.3, respectively (Figure 3b). This range of pH can contribute to improving plant nutrition and decreasing exchangeable aluminum and manganese in soil solutions. This result also indicates that the addition of 30 t MSWC ha^{-1} to acidic soil was enough to increase soil pH to adequate values.

Conversely, as expected, the calcium nitrate addition did not increase soil pH values. This could be due to the increase in soil buffering capacity. On other hand, the nitrification of ammonium-N from fertilizer (15.5% N-NO$_3$, 1.1% N-NH$_4$, and 18.6% Ca) or ammonium uptake by plants can contribute to reducing soil pH [29].

Soil EC$_e$ values in both soils increased with the rate of MSWC, supplemented or not with nitrogen addition (Figure 3c). Despite the increase, the EC$_e$ values in both soils were very low, not exceeding 0.40 dS m^{-1} (Figure 3c). This value is much lower than the value that inhibited spinach plant growth (2 dS m^{-1}) or the growth of most crop vegetables or fruit crops [30,31]. These results indicate that despite the high EC$_e$ of MSWC (5.4 dS m^{-1}), its addition to both soils, even at the highest rate (70 t), did not reach values that affect crop growth. This was also reported in soil cultivated with spiny chicory [32] and spinach [7], which was added at similar rates to compost with high EC$_e$.

The soil nitrate contents in both soils significantly increased with the rate of MSWC, but were not affected by the nitrogen addition (Figure 3d). This could be due to higher nitrogen uptake by the plants grown with MSWC plus nitrogen than those grown only with compost (Figure 5a) or the date of the last application of inorganic nitrogen (one week before spinach harvest).

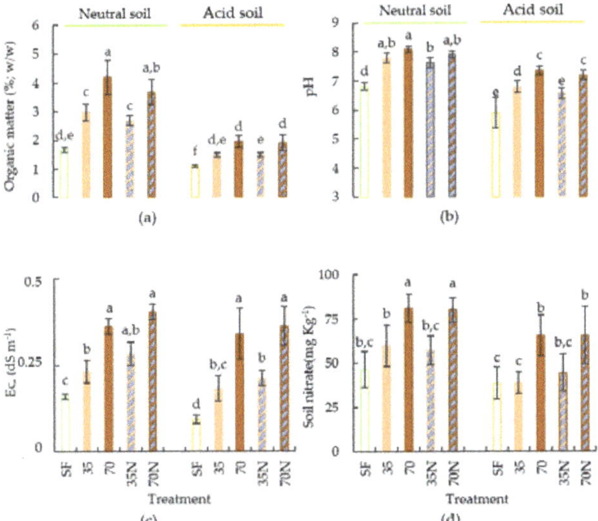

Figure 3. Effects of soil and MSWC supplemented or not with inorganic nitrogen on soil organic matter (a), pH (b), EC$_e$ (c) and nitrate (d) values. Note: FS—fertilized soil; 35—35 t MSWC ha^{-1}; 70—70 t MSWC ha^{-1}; 35 + N—35 t MSWC ha^{-1}+ 92 kg N ha^{-1}; 70 + N—70 t MSWC ha^{-1} + 92 kg N ha^{-1}. Means with different letters are significantly different at $p < 0.05$. Each bar represents the mean of six replicates, and the error bars represent ±1SE.

3.2. Plant Growth and Yield

Shoot dry weight, foliar area, and yield (fresh yield) values were not affected by the interactions between treatments. The soil significantly influenced the yield ($p < 0.001$), which was higher in the neutral soil.

Plants grown with inorganic fertilization had greater shoot dry weight values than those grown with compost or compost plus nitrogen (Table 4). Shoot dry weight values increased with inorganic nitrogen addition to compost and with the rate of compost (Table 4). Nitrogen addition to the MSWC at both rates led to an increase in shoot dry weight by $\approx$100%.

Plants grown with the highest rate of compost plus nitrogen had greater leaf area than those grown with the other treatments (Table 4).

Table 4. Effects of soil and MSWC supplemented or not with inorganic nitrogen on shoot dry weight, foliar area, and fresh yield of spinach.

MSWC Treatments	Shoot Dry Weight (g/plant)		Foliar Area (cm^2/plant)		Fresh Yield (kg m^{-2})	
	Soil		Soil		Soil	
	Neutral	Acidic	Neutral	Acidic	Neutral	Acidic
FS	1.70 a [1]	1.44 a	215.20 b	240.98 b	3.97 a	3.60 a
35	0.59 d	0.52 d	101.24 d	82.80 d	1.61 d	1.23 d
70	0.72 d	0.71 d	126.83 c	124.08 c	2.03 c	1.90 c
35 + N	1.17 c	1.13 c	254.85 a	239.72 b	3.43 b	3.05 b
75 + N	1.44 b	1.42 b	258.30 a	255.25 a	3.86 a	3.68 a

[1] Means followed by different letters within a column are significantly different ($p < 0.05$). Note: FS—fertilized soil; 35—35 t MSWC ha^{-1}; 70—70 t MSWC ha^{-1}; 35 + N—35 t MSWC ha^{-1} + 92 kg N ha^{-1}; 70 + N—70 t MSWC ha^{-1} + 92 kg N ha^{-1}.

Spinach fresh yield significantly increased with the rate of MSWC supplemented or not with nitrogen addition.

The addition of inorganic nitrogen to the 35 and 70 t of MSWC led to average increases in fresh yield of 147 and 90%, respectively. The increases in fresh yield due to inorganic nitrogen at both rates of MSWC and for both soils were similar.

The combined addition of inorganic nitrogen (43% of the inorganic nitrogen applied to fertilized soil) with the highest rate of compost addition increased the yield from the neutral (3.86 kg m^{-2}) and acidic soils (3.68 kg m^{-2}) to similar values to those obtained with inorganic fertilization (Table 4).

3.3. Shoot Nutrient Concentration and Uptake

Shoot N, P, K, Ca, and Mg contents were not influenced by the interactions between the treatments. Such behavior was also reported by Papafilippaki [32] in spiny chicory grown in soils with different MSWC rates. Shoot macronutrient concentrations, except for shoot N and Ca, were significantly affected by the soil, being higher in neutral soil.

The rate of MSWC combined or not with inorganic N had no significant influence on shoot N concentration (Figure 4a). Leaf N concentrations in plants grew only when MSWC was below the recommended level (2 to 4%) [33]. Calcium nitrate addition to compost increased shoot N concentrations (averaging 65% in both soils). Shoot N uptake followed the same trend, however in both soils the values were lower than those in plants grown in fertilized soil (Figure 5a).

Plants grown only with compost had higher shoot P concentrations than those grown only with conventional fertilization and compost plus nitrogen (Figure 4b). This increase has also been mentioned by other authors [7,14,34]. The addition of MSWC to the soil can increase shoot P uptake directly by supplying P and indirectly due to the addition of the humic substances (humic acids, fulvic acids, humins, etc.) and changes in the pH. Moreover, MSWC can also increase phosphorus uptake, since it can increase its diffusive flux and availability due to increases in soil moisture content (Figure 2) and microbial activity [35,36]. Calcium nitrate addition to compost significantly decreased shoot P concentrations (Figure 4b). Despite this, in both soils, the shoot P uptake of plants grown with the highest rate of compost plus calcium nitrate was equal to plants grown in the fertilized soil (Figure 5b). The shoot P concentration was within the required range (0.3–0.4%) [33]. This indicated that the combined application of the highest rate of compost plus nitrogen can contribute to reducing inorganic P application.

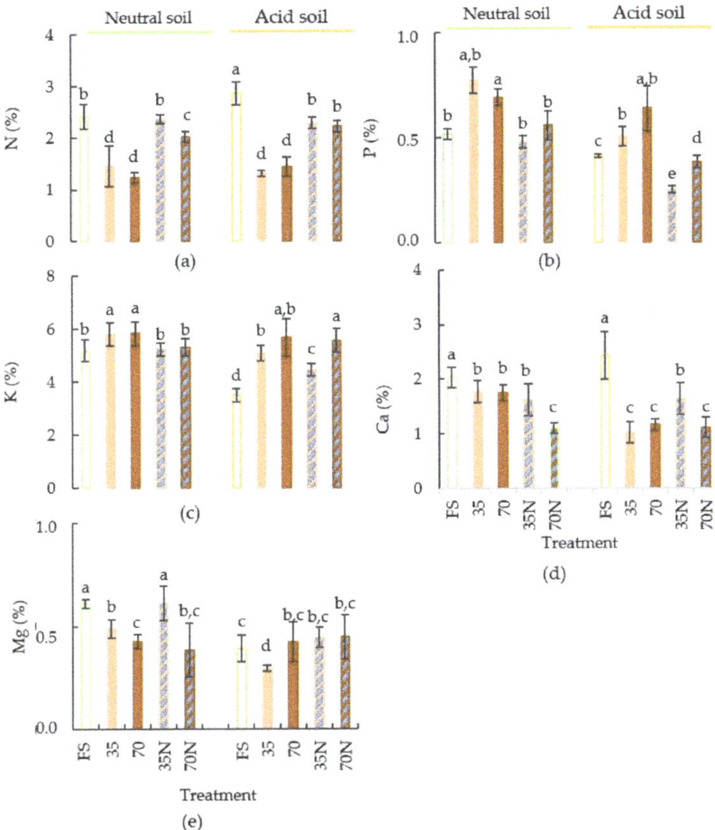

Figure 4. Effects of soil and MSWC supplemented or not with inorganic nitrogen on shoot N (**a**), P (**b**), K (**c**), Ca (**d**), and Mg (**e**) concentrations. Note: FS—fertilized soil; 35—35 t MSWC ha^{-1}; 70—70 t MSWC ha^{-1}; 35 + N—35 t MSWC ha^{-1}+ 92 kg N ha^{-1}; 70 + N—70 t MSWC ha^{-1} + 92 kg N ha^{-1}. Means with different letters are significantly different at $p < 0.05$. Each bar represents the mean of six replicates, while the error bars represent ±1SE.

Plants grown only with compost at both rates and in both soils had higher shoot K concentrations (averaging 0.12% and 0.53% in neutral and acidic soils, respectively) than those plants grown using conventional fertilization (Figure 4c). This is consistent with other studies that also reported increases in plant tissue K contents in other crops [37–40] due to the addition of MSWC. This could be due to the fact that the K in MSWC is easily available to plants [10], since normally more than 75% of the potassium in compost is soluble [41]. In lettuce, another short-term crop similar to spinach, the use of MSWC also increased the leaf K content compared with plants grown with inorganic fertilizer [17]. Despite the difference in the K$_2$O exchangeable contents between soils (204 and 60 mg kg^{-1} in neutral and acid soils, respectively) (Table 1), shoot K uptake by the plants grown with the highest rate of compost plus inorganic N was equal (neutral soil) or higher (acidic soil) than in those plants grown in fertilized soil (Figure 5c). This shows that the addition of MSWC plus nitrogen could replace inorganic K fertilization to a significant extent.

Shoot Ca concentrations of the plants grown in fertilized soils were higher than those in the plants grown using the other treatments (Figure 4d). However, in these treatments, average shoot Ca values were equal to the lower end or within the sufficiency range (1–1.5%) [33] (Figure 4d).

Despite the high concentration of Ca in MSWC (18.5 g Ca kg^{-1} on a dry weight basis), shoot calcium concentrations did not increase with the MSWC rate (Figure 4d). In basil (*Ocimum basilicum*, L.), it was also reported [38] that the addition of compost increased the soil calcium concentration but not plant Ca uptake. Calcium nitrate addition to the compost led to a decrease in shoot Ca concentration (Figure 4d). This was due to a dilution effect, since in both soils the addition of calcium nitrate to the compost led to an increase in shoot Ca uptake (Figure 5d).

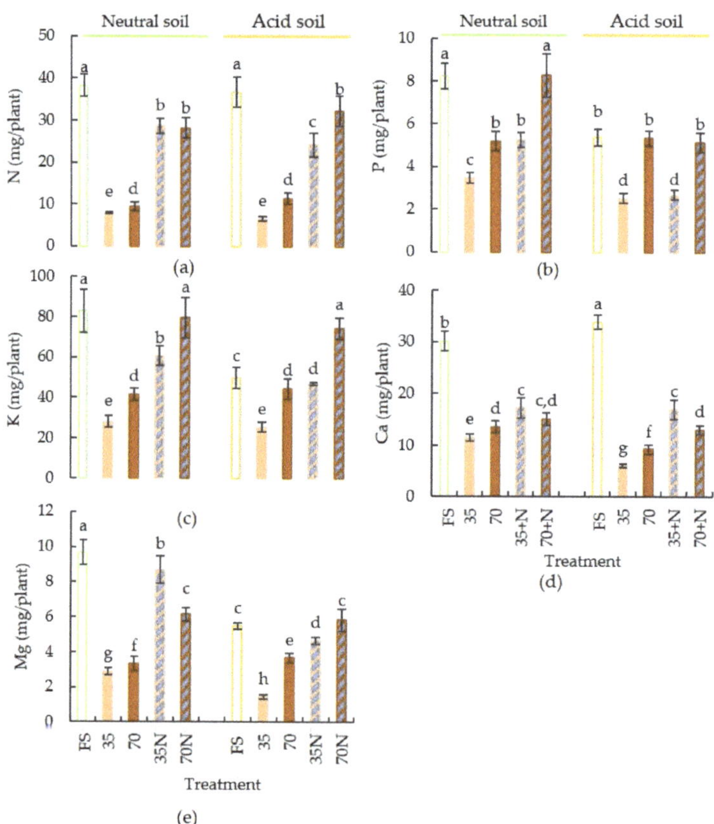

Figure 5. Effects of soil and MSWC supplemented or not with inorganic nitrogen on shoot N (**a**), P (**b**), K (**c**), Ca (**d**), and Mg (**e**) uptake rates. Note: FS—fertilized soil; 35—35 t MSWC ha-1; 70—70 t MSWC ha^{-1}; 35 + N—35 t MSWC ha^{-1}+ 92 kg N ha^{-1}; 70 + N—70 t MSWC ha^{-1} + 92 kg N ha^{-1}. Means with different letters are significantly different at $p < 0.05$. Each bar represents the mean of six replicates, while the error bars represent ±1SE.

Average shoot Mg concentrations in neutral and acidic soils ranged from 0.38 to 0.59% and from 0.29 to 0.44%, respectively (Figure 4e). These ranges of values were below or slightly higher than the lower end of the range considered to be sufficient

(0.4 to 1%) [33]. This indicates that the plants may have been subject to Mg deficiency. However, none of the plants in the treatment groups showed visual symptoms of Mg deficiency. Shoot Mg concentrations in plants grown only with compost in the neutral soil decreased, while those in the acidic soil increased (Figure 4e). In lettuce [17] and in blueberry (*Vaccinium angustifolium*, L.) [42], shoot Mg concentrations also increased with MSWC rate. The addition of nitrogen to compost in both soils did not increase shoot Mg concentrations (Figure 4), but contributed to increasing shoot Mg uptake (Figure 5e).

Shoot micronutrient concentrations, except for Fe and Cu, were significantly affected by the interactions between the treatments. Shoot Fe and Cu were not significantly affected by soil or compost treatments. Therefore, the increase of the pH did not decrease shoot Fe contents, which may have been due to the Fe complexation by humic acids [43,44] in the compost (Table 2). Shoot Fe and Cu concentrations ranged from 89 to 135.8 $\mu g\ g^{-1}$ and from 6 to 9.23 $\mu g\ g^{-1}$, respectively. These values were within the range considered to be sufficient for spinach [33].

Shoot Mn contents in both soil types were higher in plants grown in fertilized soil than those grown with the other treatments, particularly in the acidic soil (Figure 6a).

The addition of compost supplemented or not with calcium nitrate significantly decreased shoot Mn concentrations (Figure 6a), probably due to the increase in pH, while the exchangeable Mn availability decreased in the rhizosphere. Blueberry (*Vaccinium angustifolium*, L.) leaves also had lower Mn contents in samples from MSWC-treated soils when compared to control and fertilizer treatments [42]. This result indicates that the addition of MSWC is a way to eliminate or alleviate the Mn toxicity reported in this soil by [45].

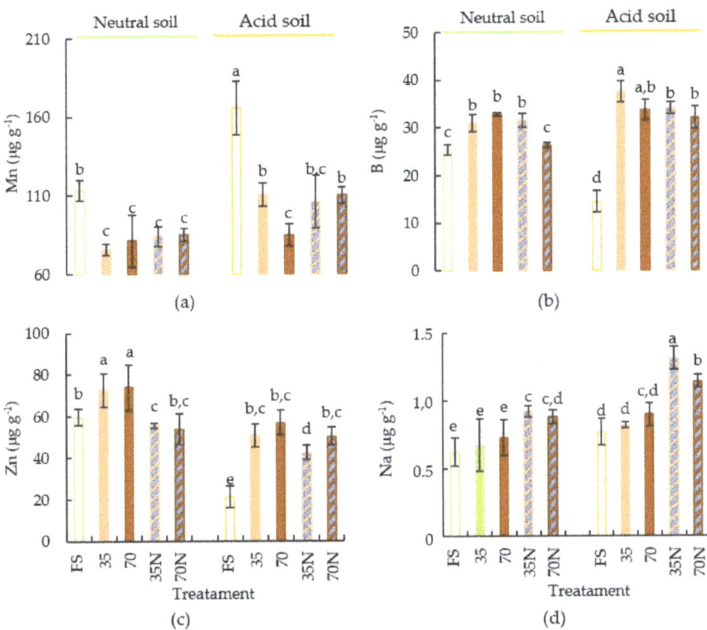

Figure 6. Effects of soil and MSWC supplemented or not with inorganic nitrogen on shoot Mn (**a**), B (**b**), Zn (**c**) and Na (**d**) concentrations. Note: FS—fertilized soil; 35—35 t MSWC ha^{-1}; 70—70 t MSWC ha^{-1}; 35 +N—35 t MSWC ha^{-1}+ 92 kg N ha^{-1}; 70 + N—70 t MSWC ha^{-1} + 92 kg N ha^{-1}. Means with different letters are significantly different at $p < 0.05$. Each bar represents the mean of six replicates, while the error bars represent ±1SE.

Despite the increase in soil pH, shoot Zn concentrations increased with the addition of MSWC, but did not increase more with increased rate of MSWC (Figure 6c). This may have been related to the increase of pH with the rate of MSWC (Figure 3b), which can contribute to reducing the availability of Zn in soil solution. The increase of shoot Zn with MSWC addition was also reported by Rajaie [46] in tomato and by Giannakis [17] in tomato and lettuce.

The addition of calcium nitrate to compost in both soils decreased shoot Zn contents (Figure 6c), which may have been due to their dilution, since shoot Zn uptake increased.

Average shoot Zn concentrations in both soil types and with both treatments ranged from 21 to 73.8 μg g^{-1} dry matter (DW). These values are lower than those that inhibit the growth of most plants (200–500 μg g^{-1} DM [47] and 100–700 μg g^{-1} DM [48,49], respectively). Indeed, average shoot Zn concentrations, except in the acidic, fertilized soil treatment (21 μg g^{-1} DW), were within the range considered to be sufficient for spinach (25–75 μg g^{-1}) [33].

Shoot B concentrations, except in the 70 + N treatment in neutral soil, were significantly higher in plants grown with MSWC supplemented or not with inorganic nitrogen than those grown in fertilized soil (Figure 6b). In these treatments, average shoot B concentrations ranged from 26.30 to 37 μg g^{-1}. These values were slightly below or within the ranges considered to be sufficient (30 to 50 μg g^{-1} DW [50] and 25 to 60 μg g^{-1} DW [33], respectively).

Despite the differences, the leaf micronutrient concentrations in the different treatments in both soils were always below toxic levels. Despite the differences, the micronutrient concentrations in the different treatments in both soils were always below toxic levels. Despite this, the low concentration of heavy metals in compost will be important in further studies to evaluate the influence of the application of MSWC on heavy metal concentrations in these soils.

Shoot Na concentrations were not affected by the interactions between treatments. Shoot Na concentrations were significantly higher in acidic soil. Shoot Na concentrations in plants grown only with compost at both rates (35 and 70 t) were not significantly different from the plants grown with conventional fertilization (Figure 6d). However, the addition of calcium nitrate to compost significantly increased leaf Na concentrations to values ranging from 0.90 and 1.3 μg g^{-1} (Figure 6d), which may have been due to an increase in the Na availability in the soil solution due to replacing Na in the soil exchange complex with Ca.

3.4. Photosynthetic Pigments

Leaf chlorophyll a and b and total chlorophyll (Chl a+b) contents in fresh weight were significantly affected by the interactions of the treatments ($p < 0.001$), indicating that the responses to the addition of MSWC supplemented or not with nitrogen differed among soils. However, Chl a contents in both soils were signifcantly higher in plants grown in fertilized soil than those grown using the other treatments (Figure 7a). In neutral soil, the Chl a decreased with MSWC supplemented or not with nitrogen. However, in acidic soil, the rate of compost had no influence on Chl a content (Figure 7a).

The Chl b content in neutral soil in plants grown with compost plus nitrogen was not significantly different from those plants grown in the fertilized soil. In acidic soil, Chl b was higher in plants grown with 35 t MSWC/ha than in those grown with the other treatments (Figure 7b).

The total chlorophyll content was highest in the neutral fertilized soil (50 mg/100 g FW) (Figure 7c). The increase in compost rate supplemented or not with nitrogen led to a decreased in total chlorophyll. This may have been due to a dilution effect, since plants grown with the highest MSWC rate had lower shoot dry weight percentages and high leaf area, or may have been due to differences in shoot nutrient concentrations or uptake (Figures 4–6). Chlorophyll synthesis is dependent on various nutrients [51], including micronutrients [52]. A nutrient deficiency strongly influences the photosynthetic apparatus structure and functions [53].

The decrease of the chlorophyll content may have also been due to the uptake of certain trace elements (metals) not measured in the present study that can negatively influence Chl a and b contents [54]. In acidic soil, the total chlorophyll content was not significantly affected by any treatment, with values ranging from 20 to 22 mg/100 g FW. These values are low compared to those recorded by Hussain (65.4 mg/100 g FW [55] and 96.2 to 301.8 mg/100 g FW) [56] and Machado (53 to 66 mg/100 g FW) [7]. Total chlorophyll was positively correlated with leaf Mg (r = 0.65 $p < 0.01$, data from both soils and treatments), and in both soils, as previously mentioned, plants grown with certain treatments may have been subject to magnesium deficiency. This indicates that in short-term crops it may be important to add some inorganic Mg to the compost to increase the leaf chlorophyll content.

Leaf carotenoid (Cc) contents were not significantly affected by the interactions between treatments or by soil type. The Cc was higher in the FS treatment (Figure 7d), this can be because of high nitrogen application. In spinach [7] and kale [57], the Cc content increased with nitrogen application. Indeed, the Cc content was positively correlated with leaf N content (r = 0.651 $p < 0.01$). The rate of compost without nitrogen supplementation in both soils did not significantly affect the Cc contents. However, Cc contents only increased significantly in plants grown with the highest compost rate supplemented with nitrogen (Figure 7d). Despite the differences in Cc levels, the values measured (ranging from an average of 22 to 40 mg/100 g FW (Figure 7d)) were within or above the value ranges reported by Borowski and Michalek (17 to 32 mg/100 FW) [58] and Machado (21.5 to 31.1 mg/100 g FW) [7].

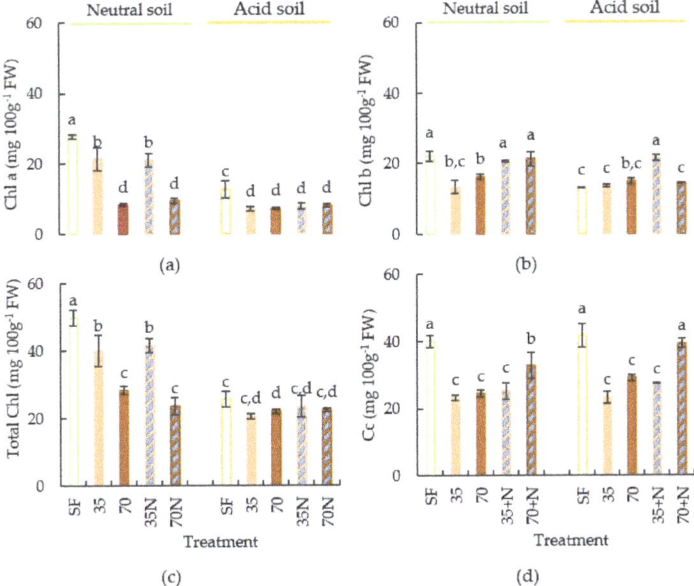

Figure 7. Effects of soil and MSWC rates supplemented or not with inorganic nitrogen on levels of photosynthetic pigments [chlorophyll a (**a**), chlorophyll b (**b**), total chlorophyll (Chl a+b) (**c**)], and carotenoids (**d**). Note: FS—fertilized soil; 35—35 t MSWC ha^{-1}; 70—70 t MSWC ha^{-1}; 35 + N—35 t MSWC ha^{-1} + 92 kg N ha^{-1}; 70 + N—70 t MSWC ha^{-1} + 92 kg N ha^{-1}. Means with different letters are significantly different at $p < 0.05$. Each bar represents the mean of six replicates, while the error bars represent ±1SE.

3.5. Nitrate

Shoot NO$_3{}^-$ concentrations were significantly influenced by the interactions between the treatments. However, in both soils and with the different treatments, shoot nitrate concentrations always below the maximum value allowed by the European Union for fresh spinach (3.5 mg g^{-1} fresh weight) [59] (Figure 8). This result indicates that the addition of

MSWC at either rate, supplemented or not with inorganic nitrogen, does not represent an issue for NO_3^- concentrations in spinach.

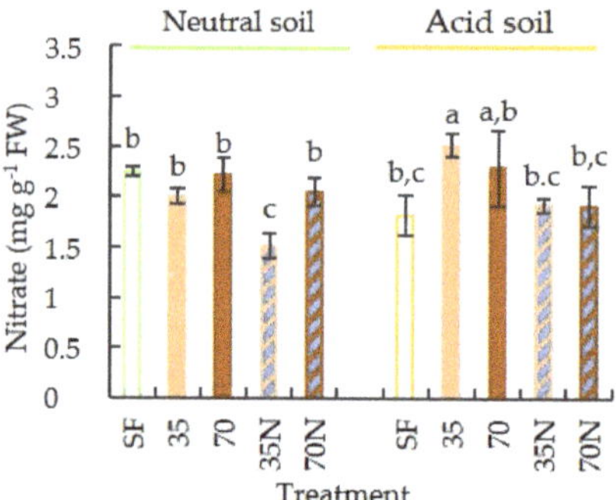

Figure 8. Effects of soil and MSWC supplemented or not with inorganic nitrogen on shoot nitrate concentrations. Note: FS—fertilized soil; 35—35 t MSW compost ha^{-1}; 70—70 t MSW compost ha^{-1}; 35 + N—35 t MSWC ha^{-1} + 92 kg N ha^{-1}; 70 + N—70 t MSWC ha^{-1} + 92 kg N ha^{-1}. Means with different letters are significantly different at $p < 0.05$. Each bar represents the mean of six replicates, while the error bars represent ±1SE.

3.6. Antioxidant Activity (DPPH)

Leaf spinach antioxidant activity (DPPH) levels were not significantly affected by the interactions between treatments or by soil type. Leaf DPPH levels in both soils, regardless of the applied rate, were lower in treatments where MSWC was applied but not combined with nitrogen (Figure 9).

Overall, leaf DPPH levels increased with the addition of nitrogen to the compost (Figure 9). This may have been due to the increase in yield, which consequently increased nutrient plant uptake, which may have led to the occurrence of nutrient deficiency, as apparently occurred with magnesium.

As apparently occurred with magnesium, the antioxidant activity increased with nutritional stress, deficiency, imbalance, and specific toxicities. For example, Mg deficiency increased the activity of antioxidant enzymes and the concentrations of antioxidant molecules in beans [60] and pepper [61].

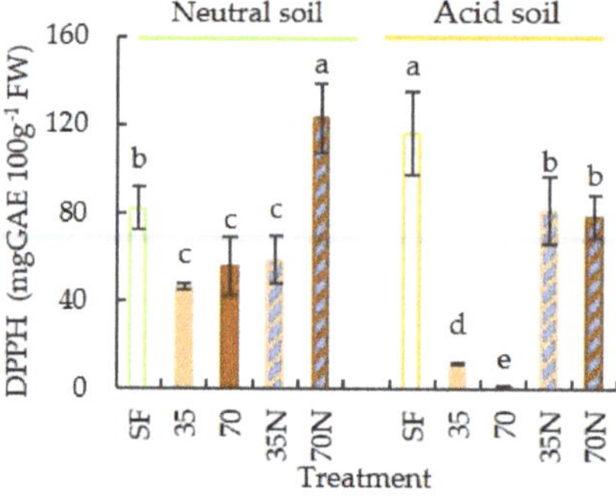

Figure 9. Effects of soil and MSWC supplemented or not with inorganic nitrogen on shoot antioxidant activity (DPPH) levels. Note: FS—fertilized soil; 35—35 t MSW compost ha^{-1}; 70—70 t MSW compost ha^{-1}; 35 + N—35 t MSWC ha^{-1} + 92 kg N ha^{-1}; 70 + N—70 t MSWC ha^{-1} + 92 kg N ha^{-1}. Means with different letters are significantly different at $p < 0.05$. Each bar represents the mean of six replicates, while the error bars represent ±1SE.

In treatments where nitrogen was applied, DPPH levels (ranging from an average of 45 to 123 mg GAE/100 g FW (Figure 9)) were similar to or above the values reported by Galani (100 mg 100 g^{-1} FW) [62] and Machado (20.54 to 31.1 mg 100 g^{-1} FW) [7].

Thus, further studies should evaluate the influence of MSWC supplemented with inorganic nitrogen and magnesium on plant growth, photosynthetic pigments, and oxidative stress markers such as ROS-scavenging enzymes (e.g., peroxidases, catalases) and secondary metabolites (e.g., phenols, ascorbate, glutathione).

4. Conclusions

The addition of MSWC to soil increased soil organic matter and pH values in both soils. Regardless of the rate of MSWC added to acidic soils, pH increased to adequate values for plant growth (close to neutral). Plant growth in both soils increased with the addition of inorganic nitrogen to the compost and with the rate of compost added. The supplementation of the highest rate of MSWC (70 t of MSWC) a fresh yield similar to those obtained in the fertilized soils, and substantially reduced the amount of inorganic nitrogen applied. Weekly addition of inorganic nitrogen to MSWC increased levels of shoot N, P, K, Ca, and Mg uptake. The addition of nitrogen to the highest MSWC rate increased shoot P and K to levels similar to those grown with inorganic fertilization treatment. MSWC addition reduced the shoot Mn concentration considerably. Regardless of the MSWC rate or addition of N, leaf tissue concentrations of Zn, Fe, Mn, and Cu did not reach toxic levels. Shoot NO3 concentrations were also lower than the maximum allowed by the European Union for fresh spinach. The supplementation of the 70 t rate of MSWC with inorganic nitrogen increased leaf antioxidant activity in both soils.

Author Contributions: R.M.A.M. conceived and designed the experiments; performed the experiments; analyzed and interpreted the data; contributed reagents, materials, analysis tools, and data; and wrote the paper. M.R. performed the experiments and analyzed the data. I.A.-P. and R.F. performed the experiments; analyzed and interpreted the data; contributed reagents, materials, analysis tools, and data; and wrote the paper. All authors have read and agreed to the published version of the manuscript.

Funding: This research was funded by the Foundation for Science and Technology (FCT) as part of Project UIDB/05183/2020.

Institutional Review Board Statement: Not applicable.

Informed Consent Statement: Not applicable.

Data Availability Statement: Not applicable.

Conflicts of Interest: The authors declare no conflict of interest.

References

1. European Communities. Sustainable Agriculture and Soil Conservation Soil Degradation Processes. 2009. Available online: https://esdac.jrc.ec.europa.eu/projects/SOCO/FactSheets/ENFactSheet-03.pdf (accessed on 18 December 2020).
2. Jones, A.; Panagos, P.; Barcelo, S.; Bouraoui, F.; Bosco, C.; Dewitte, O.; Gardi, C.; Erhard, M.; Hervás, R.; Hiederer, R.; et al. The State of Soil in Europe. A Contribution of the JRC to the European Environment Agency's Environment State and Outlook Report. (European Commission: Luxembourg). 2012. Available online: https://esdac.jrc.ec.europa.eu/ESDB_Archive/eusoils_docs/other/EUR25186.pdf (accessed on 15 January 2021).
3. Von Uexküll, H.R.; Mutert, E. Global extent, development and economic impact of acid soils. *Plant Soil* **1995**, *171*, 1–15. [CrossRef]
4. Kunhikrishnan, A.; Thangarajan, R.; Bolan, N.; Xu, Y.; Mandal, S.; Gleeson, D.; Seshadri, B.; Zaman, M.; Barton, L.; Tang, C.; et al. Functional Relationships of Soil Acidification, Liming, and Greenhouse Gas Flux. In *Advances in Agronomy*; Academic Press: Amsterdam, The Netherlands, 2016; Volume 139, pp. 1–71. [CrossRef]
5. Avillez, F.; Carvalho, M. A importância de uma gestão sustentável do solo para o crescimento da agricultura portuguesa. *Cultiv. Cad. Análise Prospetiva.* **2015**, *2*, 27–40.
6. Bian, M.; Zhou, M.; Sun, D.; Li, C. Molecular approaches unravel the mechanism of acid soil tolerance in plants. *Crop J.* **2013**, *1*, 91–104. [CrossRef]
7. Machado, R.; Alves-Pereira, I.; Lourenço, D.; Ferreira, R. Effect of organic compost and inorganic nitrogen fertigation on spinach growth, phytochemical accumulation and antioxidant activity. *Heliyon* **2020**, *6*, 05085. [CrossRef] [PubMed]

8.   Manicone, F. *Municipal Solid Waste Compost Evaluation as Possible Substitute of Mineral Fertilizers in Open Field and Con-trolled Environment Cultivation*; Universitá Degli Studi Della Basilicata: Matera, Italy, 2020; 100p.

9.   Amlinger, F.; Favoino, E.; Pollak, M.; Peyr, S.; Centemero, M.; Caima, V. *Heavy Metals and Organic Compounds from Wastes Used as Organic Fertilisers*; Study on Behalf of the European Commission, Directorate-General Environment: Hochbergstr, Austria, 2004; pp. 168–210.

10.   Hargreaves, J.; Adl, M.; Warman, P. A review of the use of composted municipal solid waste in agriculture. *Agric. Ecosyst. Environ.* **2008**, *123*, 1–14. [CrossRef]

11.   Domínguez, M.; Núñez, R.P.; Piñeiro, J.; Barral, M.T. Physicochemical and biochemical properties of an acid soil under potato culture amended with municipal solid waste compost. *Int. J. Recycl. Org. Waste Agric.* **2019**, *8*, 171–178. [CrossRef]

12.   Linden, A.V.; Brusselaers, J. *European Environment Agency, 2020 Bio-Waste in Europe—Turning Challenges into Opportunities*; Publications Office of the European Union: Luxembourg, 2020; ISSN 1977-8449.

13.   Smith, S.R. A critical review of the bioavailability and impacts of heavy metals in municipal solid waste composts compared to sewage sludge. *Environ. Int.* **2009**, *35*, 142–156. [CrossRef]

14.   Mkhabela, M.; Warman, P. The influence of municipal solid waste compost on yield, soil phosphorus availability and uptake by two vegetable crops grown in a Pugwash sandy loam soil in Nova Scotia. *Agric. Ecosyst. Environ.* **2005**, *106*, 57–67. [CrossRef]

15.   Hue, N.V. Correcting soil acidity of a highly weathered Ultisol with chicken manure and sewage sludge. *Commun. Soil Sci. Plant Anal.* **1992**, *23*, 241–264. [CrossRef]

16.   McCauley, A.; Jones, C.; Jacobsen, J. *Nutrient Management-Module no 8. Soil pH and Organic Matter*; Montana State University: Bozeman, MT, USA, 2016; pp. 1–16.

17.   Giannakis, G.V.; Kourgialas, N.N.; Paranychianakis, N.V.; Nikolaidis, N.P.; Kalogerakis, N. Effects of Municipal Solid Waste Compost on Soil Properties and Vegetables Growth. *Compos. Sci. Util.* **2014**, *22*, 116–131. [CrossRef]

18.   ICT. Atmospheric Sciences Water and Climate. 2019. Available online: http://www.ict.uevora.pt/g1/index.php/meteo-data/ (accessed on 20 June 2020).

19.   Decreto-Lei n° 103/2015—Ministério da Economia Diário da República, 1.ª série—N° 114—15 de junho de 2015. Available online: https://dre.pt/home/-/dre/67477872/details/maximized?p_auth=XCOzR77X&serie=I (accessed on 29 December 2020).

20.   Prazeres, A.O. *Comparação de Metodologias Laboratoriais Para Determinação de Azoto Nítrico e Amoniacal em Solos e águas*; Programa e Livro de Resumos do 1° Congresso Nacional e Rega e Drenagem: Beja, Portugal, 2005; Volume 59, pp. 1–12.

21.   Pribyl, D.W. A critical review of the conventional SOC to SOM conversion factor. *Geoderma* **2010**, *156*, 75–83. [CrossRef]

22.   Lastra, O.C. Derivate spectrophotometric determination of nitrate in plant tissue. *J. AOAC Int.* **2003**, *86*, 1001–1005. [CrossRef]

23.   Lichtenthaler, H.K.; Buschmann, C. Chlorophylls and Carotenoids: Measurement and Characterization by UV-VIS Spectroscopy. *Curr. Protoc. Food Anal. Chem.* **2001**, *1*, F4.3.1–F4.3.8. [CrossRef]

24.   Brand-Williams, W.; Cuvelier, M.; Berset, C. Use of a free radical method to evaluate antioxidant activity. *LWT* **1995**, *28*, 25–30. [CrossRef]

25.   Zhang, M.; Heaney, D.; Henriquez, B.; Solberg, E.; Bittner, E. A Four-Year Study on Influence of Biosolids/MSW Cocompost Application in Less Productive Soils in Alberta: Nutrient Dynamics. *Compos. Sci. Util.* **2006**, *14*, 68–80. [CrossRef]

26.   Paradelo, R.; Barral, M.T. Availability and fractionation of Cu, Pb and Zn in an acid soil from Galicia (NW Spain) amended with municipal solid waste compost. *Span. J. Soil Sci.* **2017**, *7*, 31–39. [CrossRef]

27.   García-Gil, J.; Ceppi, S.; Velasco, M.; Polo, A.; Senesi, N. Long-term effects of amendment with municipal solid waste compost on the elemental and acidic functional group composition and pH-buffer capacity of soil humic acids. *Geoderma* **2004**, *121*, 135–142. [CrossRef]

28.   Rousk, J.; Brookes, P.C.; Bååth, E. Contrasting Soil pH Effects on Fungal and Bacterial Growth Suggest Functional Redundancy in Carbon Mineralization. *Appl. Environ. Microbiol.* **2009**, *75*, 1589–1596. [CrossRef]

29.   Almutairi, K.F.; Machado, R.M.; Bryla, D.R.; Strik, B.C. Chemigation with Micronized Sulfur Rapidly Reduces Soil pH in a New Planting of Northern Highbush Blueberry. *HortScience* **2017**, *52*, 1413–1418. [CrossRef]

30.   Machado, R.M.A.; Serralheiro, R.P.; Machado, R.M.A.; Serralheiro, R.P. Soil Salinity: Effect on Vegetable Crop Growth. Management Practices to Prevent and Mitigate Soil Salinization. *Horticulturae* **2017**, *3*, 30. [CrossRef]

31.   Fruit Crops. In *Fruit Crops*; Elsevier: Amsterdam, The Netherlands, 2020; pp. 465–480.

32.   Papafilippaki, A.; Paranychianakis, N.V.; Nikolaidis, N.P. Effects of soil type and municipal solid waste compost as soil amendment on Cichorium spinosum (spiny chicory) growth. *Sci. Hortic.* **2015**, *195*, 195–205. [CrossRef]

33.   Campbell, C.R. Reference Sufficiency Ranges for Plant Analysis in the Southern Region of the United States, Southern Cooper-ative Series Bulletin. Available online: www.ncagr.gov/agronomi/saaesd/scsb394.pdf (accessed on 3 December 2020).

34.   Maftoun, M.; Moshiri, F.; Karimian, N.; Ronaghi, A.M. Effects of Two Organic Wastes in Combination with Phosphorus on Growth and Chemical Composition of Spinach and Soil Properties. *J. Plant Nutr.* **2005**, *27*, 1635–1651. [CrossRef]

35.   Barber, S.A. *Soil Nutrient Bioavailability: A Mechanistic Approach*, 2nd ed.; John Wiley & Sons: New York, NY, USA, 1995.

36.   Hinsinger, P.; Brauman, A.; Devau, N.; Gérard, F.; Jourdan, C.; Laclau, J.-P.; Le Cadre-Barthélémy, E.; Jaillard, B.; Plassard, C. Acquisition of phosphorus and other poorly mobile nutrients by roots. Where do plant nutrition models fail? *Plant Soil* **2011**, *348*, 29–61. [CrossRef]

37. Rodd, A.V.; Warman, P.R.; Hicklenton, P.; Webb, K. Comparison of N fertilizer, source-separated municipal solid waste compost and semi-solid beef manure on the nutrient concentration in boot-stage barley and wheat tissue. *Can. J. Soil Sci.* **2002**, *82*, 33–43. [CrossRef]

38. Zheljazkov, V.D.; Warman, P.R. Source-Separated Municipal Solid Waste Compost Application to Swiss Chard and Basil. *J. Environ. Qual.* **2004**, *33*, 542–552. [CrossRef]

39. Montemurro, F.; Maiorana, M.; Convertini, G.; Ferri, D. Compost organic amendments in fodder crops: Effects on yield, nitrogen utilization and soil characteristics. *Compost. Sci. Util.* **2006**, *14*, 114–123. [CrossRef]

40. Zheljazkov, V.D.; Astatkie, T.; Caldwell, C.D.; MacLeod, J.; Grimmett, M. Compost, Manure, and Gypsum Application to Timothy/Red Clover Forage. *J. Environ. Qual.* **2006**, *35*, 2410–2418. [CrossRef]

41. Ebertseder, T.; Gutser, R. Nutrition potential of biowaste composts. In *Applying Compost–Benefits and Needs*; Seminar Proceedings Federal Ministry of Agriculture, Forestry Environment and Water Management, Austria, and European Communities, Eds.; European Commission: Brussels, Belgium, 2003; pp. 117–128.

42. Warman, P.R.; Murphy, C.J.; Burnham, J.C.; Eaton, L.J. Soil and Plant Response to MSW Compost Applications on Lowbush Blueberry Fields in 2000 and 2001. *Small Fruits Rev.* **2004**, *3*, 19–31. [CrossRef]

43. Chen, Y.; Clapp, C.; Magen, H. Mechanisms of plant growth stimulation by humic substances: The role of organo-iron complexes. *Soil Sci. Plant Nutr.* **2004**, *50*, 1089–1095. [CrossRef]

44. Bocanegra, M.P.; Lobartini, J.C.; Orioli, G.A. Plant Uptake of Iron Chelated by Humic Acids of Different Molecular Weights. *Commun. Soil Sci. Plant Anal.* **2006**, *37*, 239–248. [CrossRef]

45. Carvalho, M.; Goss, M.J.; Teixeira, D. Manganese toxicity in Portuguese Cambisols derived from granitic rocks: Causes, limitations of soil analyses and possible solutions. *Rev. Ciências Agrárias* **2015**, *38*, 518–527. [CrossRef]

46. Rajaie, M.; Tavakoly, A.R. Effects of municipal waste compost and nitrogen fertilizer on growth and mineral composition of tomato. *Int. J. Recycl. Org. Waste Agric.* **2016**, *5*, 339–347. [CrossRef]

47. Mortvedt, J.J.; Cox, F.R.; Shuman, L.M.; Welch, R.M. *Micronutrients in Agriculture*; Soil Science Society of America Inc.: Madison, WI, USA, 1991. [CrossRef]

48. White, P.J. *The Use of Nutrients in Crop Plants*; Fageria, N.K., Ed.; CRC Press: Boca Raton, FL, USA, 2009; p. 430. ISBN 978-1-4200-7510-6.

49. White, P.J.; Brown, P.H. Plant nutrition for sustainable development and global health. *Ann. Bot.* **2010**, *105*, 1073–1080. [CrossRef] [PubMed]

50. Dechen, A.R.; Nachtigall, G.R. Micronutrientes. In *Nutrição Mineral de Plantas*; Fernandes, M.S., Ed.; SBCS: Viçosa, Brazil, 2006; Chapter XIII; pp. 328–352.

51. Fredeen, A.L.; Raab, T.K.; Rao, I.M.; Terry, N. Effects of phosphorus nutrition on photosynthesis in *Glycine max* (L.) Merr. *Planta* **1990**, *181*, 399–405. [CrossRef] [PubMed]

52. Marschner, H. *Marschner's Mineral Nutrition of Higher Plants*; Academic Press: New York, NY, USA, 2002.

53. Kalaji, H.M.; Oukarroum, A.; Alexandrov, V.; Kouzmanova, M.; Brestic, M.; Zivcak, M.; Samborska, I.A.; Cetner, M.D.; Allakhverdiev, S.I.; Goltsev, V. Identification of nutrient deficiency in maize and tomato plants by in vivo chlorophyll a fluorescence measurements. *Plant Physiol. Biochem.* **2014**, *81*, 16–25. [CrossRef] [PubMed]

54. Zengin, F.K.; Munzuroglu, O. Effects of some heavy metals on content of chlorophyll, proline and some antioxidant chemicals in bean (*Phaseolus vulgaris* L.) seedlings. *Acta Biol. Cracov. Bot.* **2005**, *47*, 157–164.

55. Hussain, P.R.; Suradkar, P.; Javaid, S.; Akram, H.; Parvez, S. Influence of postharvest gamma irradiation treatment on the content of bioactive compounds and antioxidant activity of fenugreek (*Trigonella foenum–graceum* L.) and spinach (*Spinacia oleracea* L.) leaves. *Innov. Food Sci. Emerg. Technol.* **2016**, *33*, 268–281. [CrossRef]

56. Batziakas, K.G.; Rivard, C.L.; Stanley, H.; Batziakas, A.G.; Pliakoni, E.D. Reducing preharvest food losses in spinach with the implementation of high tunnels. *Sci. Hortic.* **2020**, *265*, 109268. [CrossRef]

57. Kopsell, D.; Kopsell, D.; Curran-Celentano, J. Carotenoid pigments in kale are influenced by nitrogen concentration and form. *J. Sci. Food Agric.* **2007**, *87*, 900–907. [CrossRef]

58. Borowski, E.; Michałek, S. The effect of foliar nutrition of spinach (*Spinacia oleracea* L.) with magnesium salts and urea on gas exchange, leaf yield and quality. *Acta Agrobot.* **2012**, *63*, 77–86. [CrossRef]

59. Regulation (EU) n°1258/European Commission, Amending Regulation (EC) No 1881/2006 as Regards Maximum Levels for Nitrates in Foodstuffs. Official Journal of the European Union, 2011. Available online: https://eur-lex.europa.eu/legal-content/EN/TXT/PDF/?uri=CELEX:32011R1258&from=EN (accessed on 3 December 2020).

60. Cakmak, I.; Marschner, H. Magnesium Deficiency and High Light Intensity Enhance Activities of Superoxide Dismutase, Ascorbate Peroxidase, and Glutathione Reductase in Bean Leaves. *Plant Physiol.* **1992**, *98*, 1222–1227. [CrossRef] [PubMed]

61. Anza, M.; Riga, P.; Garbisu, C. Time course of antioxidant responses of Capsicum annuum subjected to a progressive magnesium deficiency. *Ann. Appl. Biol.* **2005**, *146*, 123–134. [CrossRef]

62. Galani, J.H.Y.; Patel, J.S.; Patel, N.J.; Talati, J.G. Storage of Fruits and Vegetables in Refrigerator Increases their Phenolic Acids but Decreases the Total Phenolics, Anthocyanins and Vitamin C with Subsequent Loss of their Antioxidant Capacity. *Antioxidants* **2017**, *6*, 59. [CrossRef]

 *horticulturae*

Article

Valorization of Spent Coffee Grounds, Biochar and other residues to Produce Lightweight Clay Ceramic Aggregates Suitable for Nursery Grapevine Production

Domenico Ronga [1,2,*]**, Mario Parisi** [3]**, Luisa Barbieri** [4]**, Isabella Lancellotti** [4]**, Fernanda Andreola** [4] **and Cristina Bignami** [2]

1   Centro Ricerche Produzioni Animali—CRPA S.p.A, Viale Timavo, n. 43/2, 42121 Reggio Emilia, Italy
2   Department of Life Science, Centre BIOGEST-SITEIA, University of Modena and Reggio Emilia, Via Amendola, n. 2, 42122 Reggio Emilia (RE), Italy; cristina.bignami@unimore.it
3   CREA Research Centre for Vegetable and Ornamental Crops, Via Cavalleggeri, 25, 84098 Pontecagnano Faiano, Italy; mario.parisi@crea.gov.it
4   Department of Engineering 'Enzo Ferrari', University of Modena and Reggio Emilia, Via Vivarelli 10, 41125 Modena, Italy; luisa.barbieri@unimore.it (L.B.); isabella.lancellotti@unimore.it (I.L.); andreola.fernanda@unimore.it (F.A.)
*   Correspondence: domenico.ronga@unimore.it; Tel.: +39-3396805848

Received: 4 July 2020; Accepted: 21 September 2020; Published: 23 September 2020

Abstract: The valorization of agro-industrial by-products is one of the key strategies to improve agricultural sustainability. In the present study, spent coffee grounds and biochar were used as pore forming agents in the realization of lightweight clay ceramic aggregates that were used as sustainable fertilizers, in addition to tailored glass fertilizer containing phosphorous (P) and potassium (K) and nitrogen (N) synthetic fertilizer, for nursery grapevine production. The obtained fertilizers were assessed in a pot experiment for the fertilization of bare-rooted vines. Unfertilized (T0) and fertilized plants (T1, using NPK-containing commercial fertilizer) were used as controls. Plants fertilized by spent coffee grounds and spent coffee grounds + biochar-containing lightweight aggregates and added with 30 wt% of the above-mentioned glass and N fertilizers (T2 and T3, respectively) recorded higher values of plant height, shoot diameter, leaf and node numbers. Moreover, T2 treatment induced the highest chlorophyll content, shoot and root dry weights. The present study shows that lightweight clay ceramic aggregates containing spent coffee grounds and glass and N fertilizers can be used for nursery grapevine production, in turn improving the agricultural sustainability.

Keywords: by-products; recycle; smart agriculture; sustainability; nitrogen efficiency

---

1. Introduction

The increasing request of grapevine (*Vitis vinifera* L.) planting materials, due to the expansion of global viticulture and vineyard renewal, is encouraging research studies to increase sustainability of this crop already starting from the nursery phase [1].

Nowadays, different grapevine planting materials are available in commerce. For the planting of a new vineyard, the 1-year-old dormant bare-rooted vines are the most adopted material [2]. It is produced by bench grafting at the end of winter-early spring the dormant one-bud cuttings of the scion onto hardwood cuttings of the rootstock collected during plant dormancy phase, subsequently growing the grafted vines in a greenhouse to achieve the callusing and thereafter transferring them in the open field for the growth and development of roots and shoot. Finally, after excavation and roots and shoot pruning, bare-rooted plants are commercialized.

Currently, several works have mainly focused attention on management of the vineyard in the different stages of its commercial life starting from planting [1,3–7], while little attention was paid to the previous stage, that is, nursery production. The production of planting materials plays an important role both for the establishing of new vineyards and for the replacement of lost or very weak vines or plants infected with trunk disease pathogens [8–10]. Greenhouse production of potted vines might be a successful alternative to open-field propagation, allowing also the transplanting of high-quality plant material throughout all seasons [2].

Soil and vineyard canopy management plays a key role in viticulture, impacting vegetative and reproductive development of the vines [3,11]. Fertilization provides essential nutrients to young plants, which reach a well-established root system and the full size of canopy. In the reproductive phase, a balanced supply of nutrients ensures optimal yield, berry composition and wine quality. Furthermore, biological and physical-chemical soil characteristics are improved by properly fertilization programs [12]. A shortage of nutrient availability due to unbalanced fertilization can lead to a reduction of plant efficiency and biomass production and allocation. On the other hand, an over-fertilization causes improper biomass partitioning, imbalances between vegetative and productive activity and increasing in production costs and in environmental impact of vine cultivation [13]. Nitrogen, phosphorus, potassium and magnesium are key elements of vineyard nutrition, with a fundamental role in the modulation of vegetative growth and development activities [4]. Among these elements, phosphorus is critical from an agronomic point of view due to the direct or indirect actions on radical activity, yielding potential and mitigation of soil copper excess toxicity symptoms [14,15]. Regarding the environmental aspect since phosphorus fertilization is mainly based on use of non-renewable phosphate rocks, alternative strategies are required [16].

Organic fertilizers, characterized by a gradual release of the nutrients into soil, represent a valuable source for the fertilization of important crops, especially in fruit trees showing longer vegetative and reproductive phases than herbaceous crops. Manure, green manure, digestate and compost are the most adopted organic fertilizers [17–20]. However, several biodegradable wastes can be exploited and recycled to produce eco-friendly fertilizing matrix, thus contributing to reuse of organic matter in the waste management processes in accordance with the concept of circular economy. An innovative process to produce controlled-release fertilizers was proposed by Andreola et al. [21,22], who suggested the valorization of agro-industrial and post-consumer residues as raw materials for clay ceramic materials, in particular lightweight aggregates (LWAs). LWA is the generic name for a group of aggregates characterized by a lower density than common aggregates such as natural sand, gravel, and crushed stones. For example, expanded clay, one of the several granular lightweight aggregates available in commerce, is produced by subjecting special natural clays to a thermal expansion and vitrification process at temperatures higher than 1200 °C [23]. For this material, porosity has an important role, in fact, it is fundamental to give lightness, and when used as fertilizer can impact on important agronomic performance like ability to release nutrients over time, draining, as well as retaining water [21].

In order to give porosity to the ceramic material, residues based on great high organic fraction can be considered, such as spent coffee grounds and biochar. One of the most widely consumed hot beverages worldwide is brewed coffee. According to European Coffee Report 2017–2018 [24], Europe is the second worldwide green coffee importer. According to ISTAT data [25], Italy exports roasted coffee of about 4 million equivalent green bags (60 kg bag^{-1}), and spent coffee grounds (SCGs) are generated as post-consumer products. SCGs are commonly treated as organic waste sent to composting plants or in the unsorted garbage that it is fired into incinerator power plants [26]. In the context of a circular economy, in the ceramic sector, the great availability of organic matter could be exploited for lightening an aggregate material, while the high calorific power could contribute to lower the firing temperature and the fuel consumption. Biochar is a fine-grained vegetable carbon extracted from the bottom of the gasifier. When it is added to the soil it improves its chemical-physical properties, increasing soil fertility [27] and crop yield [28], due to the strongly stable nature of organic carbon, which is not subject to degradation and mineralization. A further advantage of the storage of biochar in the ground is the reduction of CO_2 emissions in the atmosphere [29]. Indeed, the carbon sink process occurring in the soil rescues carbon dioxide from the cycle of carbon. In the last 15 years the use of biochar as a soil or substrate improver has been studied, while its application in building materials is starting to gain more attention recently. The use of biochar contributes to reducing the raw material consumption and the energy saving of the LAW firing process due to the residual carbon content present [30]. Therefore, the aim of the present study was the assessment of spent coffee grounds and biochar in the production of innovative LWAs in combination with a tailored glass-fertilizer based on cattle bone flour ash, K_2CO_3 and packaging glassy sand and their use in the sustainable production of potted grapevine planting materials.

2. Materials and Methods

2.1. Lightweight Aggregates Production and Characterization

Lightweight aggregates were obtained by powder sintering of a local (Modena, Italy) ferruginous red clay (85 wt%) with the addition of spent coffee grounds (SCGs) or spent coffee grounds (Modena, Italy) + biochar (B) (15 wt%) (Emilia Romagna Region, Italy) as poring agents to reduce both density and sintering temperature. Biochar was derived from gasification of woody biomass from river maintenance. The raw materials were dried in an oven at 105 °C for 24 h and subjected to a grinding and sieving process in order to reach a particle size <100 μm for clay and <250 μm for B and SCGs. A 30 wt% of a powdered glass (100 μm) containing the nutrients phosphorus and potassium, previously developed by the authors [21], based on cattle-bone flour ash (CBA), packaging glass cullet and K_2CO_3 and obtained by melting the mixture at 1400 °C in an electric furnace and quenching in water, was added to the clayey body in order to confer fertilizing effect to the materials. Spherical shape samples of 1.5–2.0 g of weight and a medium diameter of 1–1.5 cm were hand-pelletized by the addition of water to obtain the adequate plasticity required for shaping. Samples were dried at 105 °C for 24 h to reduce their moisture content and then subjected to a firing process in a laboratory electrical kiln (Lenton mod. AWF13/12, Hope Valley, United Kingdom) that caused sintering of the grains and changes in their density and porosity. In order to imitate the thermal shock produced in the rotary furnace, the samples were introduced and kept in the furnace at 1000 °C for 1 h and finally allowed to cool through natural convection. The samples were codified as reported: red clay (C) + pore forming agent [spent coffee ground (SCG) or biochar (B) followed by a number indicating the percentage introduced 15% + fertilizer glass (FG) followed by a number indicating the percentage of fertilizer glass added 30%].

The specimens after firing were subjected to different physical-chemical tests in order to determine their possible use in agriculture or green roofs. In particular, weight loss (WL%) at 1000 °C, 1 h, static water absorption (WA%) in cool distilled water after 24 h following UNI EN 772-21 (2011); apparent density (AD) by Envelope density analyzer Geo Pyc 1360 equipment (Micromeritics, Norcross, GA, USA); absolute or real density (RD) by Gas Helium-pycnometer Accupyc TM II 1340 (Micromeritics, USA); total porosity percentage (TP), calculated from the densities values by the equation: TP% = [(RD–AD)/RD]*100); pH according to UNI-EN 13037 (2012) standard, using a pH-meter (XS Instruments, pH 6);-USA); electrical specific conductivity (EC), according to UNI-EN 13038 (2012) standard, using an Oakton conductimeter CON6/TDS6 (OAKTON Instruments P.O. Vernon Hills, IL, USA).

Specific tests of release were performed according to both Italian and European regulations [31] to evaluate the release capacity of P and K nutrients as well as other elements that could be harmful for the environment. Distilled water and citric acid solution 2 vol.% were used as reaction media; the values were monitored in a period from 7 to 90 days to verify controlled release. The aggregates were tested in whole size, to simulate the conditions occurring in the soil. The liquid solutions derived were analyzed by inductively coupled plasma mass spectrometry (ICP-MS Agilent 7500a, 5301 Stevens Creek Blvd, Santa Clara, CA 95051, USA).

2.2. Nursery Greenhouse Experiments

The potting experiment was performed in a nursery greenhouse, located at Reggio Emilia (Italy) under controlled temperature ranging from 19 to 25 °C (day/night) and relative humidity varying from 60% to 70%. Plants were grown under long-day conditions (15 h light, 9 h dark; light intensity 280 μmol m^{-2} s^{-1}).

The scion/rootstock combination was 'Lambrusco Salamino' cultivar grafted onto 'Kober 5BB' as rootstock. Bare-rooted vines were manually transplanted (one plant *per* pot) in plastic pots of 4.5 L capacity and filled with commercial neutralized peat (Fondolinfa Universale, Linfa Spa, RE, Italy), containing 70% organic matter, 35% organic carbon, 0.6% nitrogen (N) and having pH 6.0. Pots were arranged in a completely randomized design with five replicates and manually irrigated every two days in order to ensure the water holding capacity of the substrate. Pests were controlled according to the integrated production rules of the Emilia-Romagna Region (Italy). Plants were pruned according to one-stem training system.

Fertilization was performed only at planting, administering, in total, 6 g of N, 1.4 g of P_2O_5 and 1.6 g of K_2O *per* pot. Experimental theses were: unfertilized treatment (T0); synthetic fertilization with Osmocote Topdress [22 (N)–5 (P_2O_5)–6 (K_2O) + 2 (MgO)] (T1); LWA based on spent coffee grounds in combination with glass and N fertilizers (T2); LWA based on spent coffee grounds and biochar in combination with glass and N fertilizers (T3).

T2 and T3 showed the following composition: 7.0 (P_2O_5)–5.9 (K_2O) + 1.7 (MgO) and 5.3 (P_2O_5)–5.4 (K_2O) + 1.9 (MgO), respectively.

In order to supply the expected nitrogen rate (6 g *per* pot) ammonium nitrite and urea (2.9 and 3.1 g of N, respectively) were applied in combination with LWAs.

2.3. Recorded Parameters

During the experiment, the main phenological growth stages were recorded following the BBCH-scale (BBCH = Biologische Bundesantalt, Bundessortenamt and CHemische Industrie, Germany) [32].

Growing media (GM) were weekly monitored for the different parameters: GM water content, temperature and electrical conductivity (using a sensor Teros 11/12, Meter Group, Washington, USA). At day 314 of the year (DOY) the following traits were measured: leaf chlorophyll (Chl) and flavonoid (Flav) contents, nitrogen balance index (NBI) as ratio between Chl and Flav [30], leaf anthocyanin content (Anth), basal shoot diameter, shoot height, number of leaves and nodes. Chl, Flav, NBI, and Anth were measured on the youngest fully expanded leaf by Dualex 4 Scientific (Dx4, FORCE-A, Orsay, France), an optical leaf-clip meter for non-destructive assessment of the physiological status of the plants [33]. Leaf temperature was recorded on the fully expanded leaf using an infra-red thermometer (Everest Instruments, Chino Hills, CA, USA). Moreover, parameters characterizing plant growth were recorded: shoot fresh weight of the scion (SFW), above-ground rootstock fresh weight (RSFW), root fresh weight (RTFW), total fresh weight (TFW), shoot dry weight (SDW), above ground rootstock dry weight (RSDW), root dry weight (RTDW), total dry weight (TDW). Using these parameters, the following indexes were calculated: fraction of total dry weight allocated to shoot (FTS) ((SDW/TDW)* 100), fraction of total dry weight allocated to above-ground rootstock (FTRS) ((RSDW/TDW)* 100), and fraction of total dry weight allocated to root (FTR) ((RTDW/TDW)* 100).

In order to compare N use across the treatments, the nitrogen efficiency index was calculated from the measured plant total dry matter (g pot^{-1}) and the amount of N applied (g pot^{-1}) and was expressed as g dry biomass g^{-1} N [34].

2.4. Data Analysis

All recorded data were analyzed using GenStat 17th software (VSN International, Hemel Hempstead, UK) for analysis of variance (ANOVA). Significant means were separated by Duncan's test at $p \leq 0.05$.

3. Results and Discussion

To produce grapevine planting materials for new vineyards or to replace diseased plants, high-standard quality attributes (such as good vigor, absence of defects, well healed graft union, in addition to the guarantees of satisfactory phytosanitary status and genetic identity of rootstock and scion), can be obtained under greenhouse conditions [35]. Nowadays the environmental sustainability of orchards or vineyards must necessarily consider the pre-planting phases, i.e., the propagation phases of the plant material (i.e., proper recycling of any plastic residues or use of biodegradable pots, low impacting irrigation, disease control and fertilization methods). For the latter agronomic practice, researchers are called to find alternative fertilizers for nursery phase ensuring good plant development and low environmental impact. Considering some previous evidences of the positive effects of spent coffee grounds and biochar on growth of horticultural and woody species [2,12,13,20,36], in the present study SCGs and B were valorized to obtain new LWAs which were for the first time tested for the fertilization of bare-rooted vines in a nursery greenhouse.

3.1. Lightweight Aggregates Characteristics

All specimens prepared using red clay together with spent coffee ground (CSCG) and mixed with biochar (CBSCG) showed good workability and a good final aggregation. Unfired samples containing biochar had a darker color but after firing it disappeared obtaining aggregates with a classic red color. The idea of mixing biochar and spent coffee ground as poring agent derives from the natural alkaline pH of the biochar (11.8) and acid pH of SCGs (5.5). Aggregates prepared with only biochar as a pore-forming agent had an alkaline pH (9.3) out of the limit. The aggregates to be used in the soil must have a pH value within the optimal range of plant comfort (6–8) and EC less than 2 mS/cm. Considering these premises, it is important to evaluate the properties of LWAs functionalized with 30 wt% of fertilizing glass (FG). Table 1 shows the results of the physical and chemical characterization.

Table 1. Physical and chemical parameters of the lightweight aggregates (LWAs).

Physical Properties	CSCG15FG30	CBSCG15FG30
Weight Loss (%)	16.73	17.82
Water Absorption (%)	14.29	13.48
App. Density (kg/m^3)	1490	1400
Calculated Total Porosity (%)	42.21	45.35
Chemical Properties	**CSCG15FG30**	**CBSCG15FG30**
pH	7.24	8.31
Electrical Conductivity (dS m^{-1})	0.24	0.36

The physical characterization permitted us to classify the specimens functionalized with 30 wt% of fertilizing glass as LWAs, because their apparent density values are lower than 2000 kg m^{-3} (UNI-EN 13055–1:2003) and their porosity higher than 40%.

pH and electrical conductivity are the most important parameters to check for an agronomic application. As can be seen in Table 1, the CSCG15FG30 shows results in line with the soil guidelines above reported values while the pH of CBSCG15FG30 is slightly higher.

By the citric acid test, the release of the main nutrients P and K was monitored until 90 days. LWAs prepared with only SCG as a pore-forming agent showed an increase of P% release from 30 to 90% while those prepared using the mix (SCG + B) as a pore-forming agent highlighted a P% release from 30 to 82%. The K% release observed was lower than P% for the two compositions analyzed. For CSCG15FG30, the values increased from 5% (7 days) to 17% in 90 days while a higher release was observed for CBSCG15FG30 from 8% (7 days) to 24% in 90 days.

These differences are putatively due to the capacity of biochar to adsorb nutrients like P and K as already suggested by Carrey et al. in 2005 [37]. However, further studies are needed to corroborate this hypothesis. In addition, regarding the use of the proposed innovative fertilizers and their possible implication for soil-plant relationship, it is well known that nutrient availability of P and K depends on soil pH and on the interaction between soil microbiota and rhizosphere [38].

3.2. Bare-Rooted Vines Production

The main parameters recorded for the growing media were reported in Table 2. Peat fertilized with LWA containing spent coffee grounds and biochar (T3) showed the highest values of growing media water content, temperature and electrical conductivity (+33%, +1% and +9%, on average). These three parameters are strictly related with the plant growth and its development. In fact, soil/growing media temperature and water availability play a key role in the growth and development of roots and the vegetative growth of the vineyard canopy. During the productive phase of the vineyard, to obtain high-quality grapes, an optimal soil water deficit pattern is required [36]. On the other hand, young vines in the nursery and after planting need a high water availability, allowing maximum growth to hasten the development of the vineyard canopy [39]. Furthermore, soil temperature is positively related with root growth and development. Indeed, root growth starts above a threshold of 6–7 °C and increases with temperature with an optimum occurring at around 30 °C [40]. In bare-rooted vines after planting the availability of carbohydrate reserves is presumably quite low, and roots, shoots and leaves may face competition for photoassimilates. Warmer soil increases the mobilization of root reserves and the shoot biomass and may enhance root growth through a higher rate of carbohydrate reserve catabolism and a consequent greater availability of energy and carbon, according to Clarke et al. [41]. Together with the good water availability, a warmer soil also enhances the uptake of nitrogen, phosphorus, potassium and other nutrients [42]. These considerations are consistent with the best growth observed in T3-treated vines and with higher values of root and shoot weights, plant height, number of nodes and shoot diameter of T2-treatment than the control (T0) (Tables 3 and 4).

Table 2. Mean physical parameters of the growing media recorded in the compared thesis during the trial.

Treatment	Growing Media Water Content ($m^3\ m^{-3}$)		Growing Media Temperature (°C)		Growing Media EC ($dS\ m^{-1}$)	
T0	0.18	b	23.48	b	0.24	a
T1	0.15	c	23.46	b	0.15	c
T2	0.18	b	24.08	a	0.22	b
T3	0.21	a	24.12	a	0.24	a

T0 = unfertilized pot; T1 = fertilized pot with commercial NPK fertilizer (Osmocote Topdress); T2 = fertilized pot with LWA based on spent coffee ground, glass and N fertilizers; T3 = fertilized pot with LWA based on spent coffee grounds + biochar, glass and N fertilizers; EC = electrical conductivity. Means followed by the same letter do not significantly differ at $p \leq 0.05$.

Regarding soil electrical conductivity, by the effect of fertilization [43], Abad et al. [44] reported 0.5 dS m^{-1} as the threshold to achieve the highest plant growth. In our experiment a lower EC was found for T1 and T2 treatments than T0; moreover, the values detected for the overall thesis never reached the threshold indicated by Abad et al. [44].

Morpho-physiological parameters of bare-rooted vines were affected by LWA applications which, induced for the overall measured traits a better performance respect to commercial fertilizer applications (Table 3). At 314 DOY, corresponding to the "Principal growth stage 9: Senescence" of the BBCH-scale, T2- and T3-fertilized plants displayed higher values of plant height, node number and shoot diameter than T1-fertilized or T0-unfertilized ones (+9%, +21% and +22% of T2 and T3 respect to T0). Within stage 9, the degree of leaf fall ranged from 93 "beginning of leaf-fall" for the fertilized theses to 97 "end of leaf-fall" for the unfertilized one, indicating a longer growth duration for fertilized plants. Moreover, the highest value of leaf number (16 *per* plant) was detected in plants fertilized with LWAs containing spent coffee grounds (T2). Regarding the physiological parameters, T2- and T3-fertilized plants reported the highest values of leaf chlorophyll and flavonoid contents, NBI and leaf temperature. Moreover, for the same treatments the lowest value of leaf anthocyanins content was measured (on average, +19%, +3%, +16%, +0.4% and -24%, respectively).

Table 3. Morphological (A) and Physiological (B) parameters recorded on bare-rooted vines measured at 314 day of the year (DOY).

(A) Treatment	Plant Height (cm)		Leaf Number (No.)		Node Number (No.)		Shoot Diameter (mm)	
T0	80.00	d	3.00	d	13.00	c	3.71	c
T1	83.50	c	10.00	c	14.00	b	3.85	b
T2	86.00	b	14.00	a	16.00	a	4.38	a
T3	89.00	a	11.67	b	5.33	a	4.65	a

(B) Treatment	Chl (μg cm^{-2})		Flav (μg cm^{-2})		Anth (μg cm^{-2})		NBI (-)		Leaf Temperature (°C)	
T0	35.80	c	0.89	b	0.24	b	40.18	b	24.18	a
T1	38.95	b	0.88	bc	0.24	b	44.06	a	23.77	b
T2	22.10	d	0.84	c	0.34	a	26.25	c	24.78	a
T3	41.20	a	0.91	a	0.19	c	45.22	a	24.37	a

T0 = unfertilized pot; T1 = fertilized pot with commercial NPK fertilizer (Osmocote Topdress); T2 = fertilized pot with LWA based on spent coffee ground, glass and N fertilizers; T3 = fertilized pot with LWA based on spent coffee grounds + biochar, glass and N fertilizers; leaf chlorophyll content (Chl); leaf flavonoid content (Flav); leaf anthocyanins (Anth); nitrogen balance index (NBI). Parameter without unit of measure (-). Means followed by the same letter do not significantly differ at $p \leq 0.05$.

Table 4. Agronomic parameters recorded on bare-rooted vines.

(A) Treatment	Root Fresh Weight (g plant^{-1})		Rootstock Fresh Weight (g plant^{-1})		Shoot Fresh Weight (g plant^{-1})		Total Fresh Weight (g plant^{-1})	
T0	38.42	c	29.90	c	7.17	c	75.48	c
T1	31.67	d	43.17	a	10.00	b	84.83	b
T2	39.83	b	37.00	b	11.50	b	88.33	b
T3	44.17	a	38.50	b	12.83	a	95.50	a

(B) Treatment	Root Dry Weight (g plant^{-1})		Rootstock Dry Weight (g plant^{-1})		Shoot Dry Weight (g plant^{-1})		Total Dry Weight (g plant^{-1})	
T0	13.28	c	17.81	c	2.46	d	33.55	c
T1	15.57	b	26.90	a	3.85	c	46.32	a
T2	14.33	b	25.88	a	4.08	b	44.29	b
T3	19.13	a	23.45	b	4.40	a	46.97	a

T0 = unfertilized pot; T1 = fertilized pot with commercial NPK fertilizer (Osmocote Topdress); T2 = fertilized pot with LWA based on spent coffee ground, glass and N fertilizers; T3 = fertilized pot with LWA based on spent coffee grounds + biochar, glass and N fertilizers; (A) = biomass fresh weight; (B) = biomass dry weight; means followed by the same letter do not significantly differ at $p \leq 0.05$.

Our results were in agreement with those of previous research reporting, for different species, improvement in plant development under low supplies of spent coffee ground or biochar [45–48]. The increased Chl and NBI values for T3 treatment suggested a higher rate of leaf photosynthetic activity, assimilation and use (protein synthesis) of mineral nitrogen from the substrate. These results were in agreement with those reported by Ronga et al. [2,45] and Bozzolo et al. [12]. Leaf chlorophyll, flavonoid, and anthocyanin content are suggested as indices of plant physiological status and are associated with the N uptake and leaf photosynthetic activity [33]. Furthermore, Chalker-Scott [49] suggested that high leaf Anth levels are related with an improvement of tolerance to abiotic and biotic stresses. Interestingly, T2-fertilized plants showed the highest values of Anth. T2 treatment showed lower values of Chl and higher of Flav than T0, due to higher values of all the measured morphological parameters. In fact, in the investigated treatments, the same amount of nutrients was used, hence plants with a higher development can result in a dilution of the macro- and micronutrients that are involved in the physiological activities and related parameters like Chl content [33]. On the other hand, biochar application can allow a slow and constant release of the nutrients in the soil/growing media for plant nutrition as previously reported by Zhao et al. [50] and Gwenzi et al. [51]. In addition, this slowing release can allow a higher plant physiological activity as observed in the present study in T3-fertilized plants. Hence, our results suggested that LWA based on spent coffee ground and biochar may be used to improve plant physiological performance (increasing in stress resistances), since supplying of LWA based on biochar increased leaf photosynthetic activity in grapevine nursery management.

The results regarding the main agronomic parameters were shown in Table 4. Among the investigated treatments, bare-rooted vines fertilized with T3 showed the highest values of root, shoot and total fresh weights (+15%, +24% and +11%, on average), while the highest value of RSFW (+16%, on average) was measured for T1-fertilized plants (Table 4A). A similar trend was also noticed for the biomass dry weight. Indeed, plants fertilized with LWA based on spent coffee ground and biochar (T3) showed the highest values of root, shoot and total dry weights (+23%, +19% and +10%, on average), while T1-fertilized plants showed the highest value of RSDW (+12%, on average). In any case, this value did not significantly differ from T2 treatments, as well as total dry weight of T3 and T2 treatments was statistically comparable among them (46.3 and 27.0 g plant^{-1}, respectively).

Results regarding the improved biomass production using LWA based on spent coffee ground and biochar were in accordance with those reported by Setti et al. [52], Raviv et al. [53] and Ronga et al. [2,45], who suggested that composts derived from several by-products like spent coffee grounds and biochar positively affected the growth of potted plants grown in greenhouses. Moreover, different works reported increases in grapevine root biomass under different soil amendments like biochar, compost and digestate [2,12,54,55]. The increased biomass was putatively due to the availability of some growth-promoting compounds in the LWAs both based on spent coffee grounds and biochar as suggested by different researchers who used the same fertilizers based on agro-industrial by-products [2,56,57]. Finally, vines showing high total dry weight can better overcome transplanting stresses [58] and this improved plant physiological state represents a very important goal for nursery growers.

Figure 1 shows the biomass allocation to the different organs: root, above-ground part of the rootstock and shoot. T3-fertilized plants displayed the highest values of FTR (+11% and +4%, on average), while the highest values of biomass allocated to rootstock (FTRS) were observed in T1- and T2-fertilized plants (+5%, on average,). As suggested by Ericsson et al. [59], changes in the source-sink organ relationship can impact on the growth of the fruit-tree. Indeed, the same authors demonstrated that a high shoot development can negatively impact on the root growth. Anyway, this behavior was also partially highlighted in T2 treatment; instead, in our work, enhanced percentages of biomass allocation to shoot did not result in reduction of partitioning to the roots. The differences reported between our results and ones reported by Ericsson et al. [59] were probably ascribed to the use of SCGs and biochar in the present study. In fact, SCGs and biochar can stimulate the growth of the different plant organs as already reported in other studies [20,60]. This physiological behavior is much appreciated by vine-growers since they can dispose of planting material with a good development of both root and shoot ensuring early and faster plant establishment after transplanting, due to the good availability of accumulated reserves.

Figure 2 reports nitrogen efficiency, representing an important parameter to assess the effect of the nitrogen fertilization on the plant growth [61,62]. T3-treated plants, as well as plant fertilized with Osmocote Dropress, showed similar nitrogen efficiency values (7.8 g of biomass produced *per* each g of N applied). The performance of LWAs based on spent coffee grounds was improved with addition of biochar. Our results confirmed that this kind of substrate acts synergistically on plant growth, similarly to the results of previous studies on the same fruit tree [2,12] and on tomato [20]. In fact, in the fertilization, the use of SCGs and biochar can allow a higher plant nutrient uptake as observed in the present study and in the already published study [2,12,20].

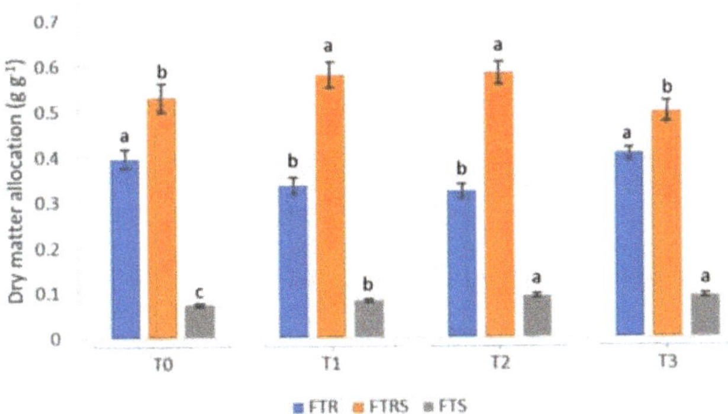

Figure 1. Dry matter allocation to different part of the bare-rooted vines in the compared thesis. T0 = unfertilized pot; T1 = fertilized pot with commercial NPK fertilizer (Osmocote Topdress); T2 = fertilized pot with LWA based on spent coffee ground, glass and N fertilizers; T3 = fertilized pot with LWA based on spent coffee grounds + biochar, glass and N fertilizers; FTS = fraction of total dry weight allocated to shoot; FTRS = fraction of total dry weight allocated to above-ground rootstock); FTR= fraction of total dry weight allocated to root Bars indicate standard deviations. Means followed by the same letter do not significantly differ at $p \leq 0.05$.

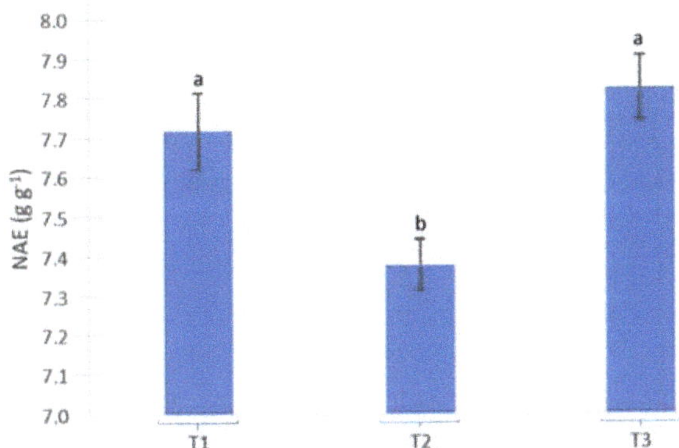

Figure 2. Comparison of nitrogen efficiency in the three fertilized theses. T1 = fertilized pot with commercial NPK fertilizer (Osmocote Topdress); T2 = fertilized pot with LWA based on spent coffee ground, glass and N fertilizer; T3 = fertilized pot with LWA based on spent coffee grounds + biochar, glass and N fertilizers; NAE = nitrogen efficiency; bars indicate standard deviations. Means followed by the same letter do not significantly differ at $p < 0.05$.

Finally, the obtained results were broadly in accordance with those of several previous studies, which reported increases in plant biomass production as effect of agro-industrial by-products, such as compost, digestate, and biochar adopted as fertilizers [2,12,18]. Increases in plant biomass were obtained by fertilizing with agro-industrial by-products and were related to the content of humic substances and some compounds able to enhance the plant growth [63]. Finally, Jindo et al. [64] reported that plant biochemical activity is improved by plant hormone-like promoters.

4. Conclusions

Results obtained in the present study suggested that spent coffee grounds and biochar represent useful components, due to their pore-forming role, for producing the matrix of lightweight aggregates.

To our knowledge, this was the first study investigating the suitability of spent coffee grounds, biochar and other residues in the form of glass in innovative slow-release fertilizer production. These recycled matrices fit well in with sustainable agriculture management and in circular economy contests. LWAs based on spent coffee grounds and biochar, when functionalized by the addition of 30 wt% of glass fertilizer, showed good agronomic performances comparable to those of synthetic fertilizer in the production of grapevine planting materials. LWAs here studied also proved to be able to reduce the needs to use phosphate and potassium rocks in nursery management phase. The positive effects on the physical characteristics of the substrate, especially on its water retention capacity, provided an interesting opportunity to save irrigation water. Finally, bare-rooted vines grown using LWAs resulted in better agronomic performances and in improved plant physiological status. However, further research is needed to assess the effect of LWAs in nursery greenhouse production of other fruit planting materials as well as in open field vine nursery production.

Author Contributions: Conceptualization, D.R. and C.B.; methodology, D.R., M.P. and C.P.; investigation, D.R., M.P., L.B., F.A., I.L. and C.B.; resources, C.B.; data curation, D.R. and C.B.; writing—original draft preparation, D.R.; writing—review and editing, D.R., M.P., F.A., L.B., I.L. and C.B.; supervision, C.B.; project administration, C.B.; funding acquisition, C.B. All authors have read and agreed to the published version of the manuscript.

Funding: This research was funded by the Fondi di Ateneo per la Ricerca di UNIMORE, FAR 2017 "La valorizzazione degli scarti agroindustriali tra diritto e scienza: processi innovativi dalla sperimentazione all'industrializzazione nel contesto legale".

Acknowledgments: The authors acknowledge and are grateful to Giovanni Grazzi (Foliae) for some of the materials used in this study.

Conflicts of Interest: The authors declare no conflict of interest.

References

1. Reynolds, A.G. *Managing Wine Quality: Viticulture and Wine Quality*; Woodhead Publishing: Cambridge, UK, 2010.
2. Ronga, D.; Francia, E.; Allesina, G.; Pedrazzi, S.; Zaccardelli, M.; Pane, C.; Tava, A.; Bignami, C. Valorization of vineyard by-products to obtain composted digestate and biochar suitable for nursery grapevine (*Vitis vinifera* L.) production. *Agronomy* **2019**, *9*, 420. [CrossRef]
3. Tesic, D.; Keller, M.; Hutton, R.J. Influence of vineyard floor management practices on grapevine vegetative growth, yield, and fruit composition. *Am. J. Enol. Vitic.* **2007**, *58*, 1–11.
4. Arrobas, M.; Ferreira, I.Q.; Freitas, S.; Verdial, J.; Rodrigues, M.Â. Guidelines for fertilizer use in vineyards based on nutrient content of grapevine parts. *Sci. Hortic.* **2014**, *172*, 191–198. [CrossRef]
5. Palliotti, A.; Tombesi, S.; Silvestroni, O.; Lanari, V.; Gatti, M.; Poni, S. Changes in vineyard establishment and canopy management urged by earlier climate-related grape ripening: A review. *Sci. Hortic.* **2014**, *178*, 43–54. [CrossRef]
6. Eldon, J.; Gershenson, A. Effects of Cultivation and Alternative Vineyard Management Practices on Soil Carbon Storage in Diverse Mediterranean Landscapes: A Review of the Literature. *Agroecol. Sustain. Food Syst.* **2015**, *39*, 516–550. [CrossRef]
7. Medrano, H.; Tomás, M.; Martorell, S.; Bota, J. Improving water use efficiency of vineyards in semi-arid regions. A review. *Agron. Sustain. Dev.* **2015**, *35*, 499–517. [CrossRef]
8. Borsellino, V.; Galati, A.; Schimmenti, E. Survey on the innovation in the Sicilian grapevine nurseries. *J. Wine Res.* **2012**, *23*, 1–13. [CrossRef]
9. Waite, H.; May, P.; Bossinger, G. Variations in phytosanitary and other management practices in Australian grapevine nurseries. *Phytopathol. Mediterr.* **2013**, *52*, 369–379.
10. Whitelaw-Weckert, M.A.; Rahman, L.; Appleby, L.M.; Hall, A.; Clark, A.C.; Waite, H.; Hardie, W.J. Co-infection by B otryosphaeriaceae and I lyonectria spp. fungi during propagation causes decline of young grafted grapevines. *Plant Pathol.* **2013**, *62*, 1226–1237. [CrossRef]
11. Salomé, C.; Coll, P.; Lardo, E.; Metay, A.; Villenave, C.; Marsden, C.; Blanchart, E.; Hinsinger, P.; Le Cadre, E. The soil quality concept as a framework to assess management practices in vulnerable agroecosystems: A case study in Mediterranean vineyards. *Ecol. Indic.* **2016**, *61*, 456–465. [CrossRef]

12. Bozzolo, A.; Pizzeghello, D.; Cardinali, A.; Francioso, O.; Nardi, S. Effects of moderate and high rates of biochar and compost on grapevine growth in a greenhouse experiment. *AIMS Agric. Food* **2017**, *2*, 113–128. [CrossRef]

13. Ciesielczuk, T.; Rosik-Dulewska, C.; Wiśniewska, E. Possibilities of coffee spent ground use as a slow action organo-mineral fertilizer. *Rocz. Ochr. Srodowiska* **2015**, *17*, 422–437.

14. Stanley, G.; Matthews, M.A. The influence of Phosphorus availability, scion, and rootstock on grapevine shoot growth, leaf area, and petiole Phosphorus concentration. *Am. J. Enol. Vitic.* **1996**, *47*, 217–224.

15. Baldi, E.; Miotto, A.; Ceretta, C.A.; Brunetto, G.; Muzzi, E.; Sorrenti, G.; Quartieri, M. Soil application of P can mitigate the copper toxicity in grapevine: Physiological implications. *Sci. Hortic.* **2018**, *238*, 400–407. [CrossRef]

16. Gautier, A.; Cookson, S.J.; Hevin, C.; Vivin, P.; Lauvergeat, V.; Mollier, A. Phosphorus acquisition efficiency and phosphorus remobilization mediate genotype-specific differences in shoot phosphorus content in grapevine. *Tree Physiol.* **2018**, *38*, 1742–1751. [CrossRef]

17. Ronga, D.; Mantovi, P.; Pacchioli, M.T.; Pulvirenti, A.; Bigi, F.; Allesina, G.; Pedrazzi, S.; Dal Prà, A. Combined Effects of Dewatering, Composting and Pelleting to Valorize and Delocalize Livestock Manure, Improving Agricultural Sustainability. *Agronomy* **2020**, *10*, 661. [CrossRef]

18. Bortolini, S.; Macavei, L.I.; Saadoun, J.H.; Foca, G.; Ulrici, A.; Bernini, F.; Malferrari, D.; Setti, L.; Ronga, D.; Maistrello, L. *Hermetia illucens* (L.) larvae as chicken manure management tool for circular economy. *J. Clean. Prod.* **2020**, *262*, 121289. [CrossRef]

19. Ronga, D.; Pellati, F.; Brighenti, V.; Laudicella, K.; Laviano, L.; Fedailaine, M.; Benvenuti, S.; Pecchioni, N.; Francia, E. Testing the influence of digestate from biogas on growth and volatile compounds of basil (*Ocimum basilicum* L.) and peppermint (*Mentha x piperita* L.) in hydroponics. *J. Appl. Res. Med. Aromat. Plants* **2018**, *11*, 18–26. [CrossRef]

20. Ronga, D.; Caradonia, F.; Parisi, M.; Bezzi, G.; Parisi, B.; Allesina, G.; Pedrazzi, S.; Francia, E. Using digestate and biochar as fertilizers to improve processing tomato production sustainability. *Agronomy* **2020**, *10*, 138. [CrossRef]

21. Andreola, F.; Borghi, A.; Pedrazzi, S.; Allesina, G.; Tartarini, P.; Lancellotti, I.; Barbieri, L. Spent Coffee Grounds in the Production of Lightweight Clay Ceramic Aggregates in View of Urban and Agricultural Sustainable Development. *Materials* **2019**, *12*, 3581. [CrossRef]

22. Andreola, F.; Lancellotti, I.; Manfredini, T.; Barbieri, L. The circular economy of agro and post-consumer residues as raw materials for sustainable ceramics. *Int. J. Appl. Ceram. Technol.* **2020**, *17*, 22–31. [CrossRef]

23. Andreola, F.; Barbieri, L.; Lancellotti, I.; Pozzi, P.; Tartarini, P.; Vezzali, V.; Allesina, G.; Pedrazzi, S. *Patent Pending 102018000009844, Procedimento per Utilizzare Char da Gassificazione e/o Pirolisi con Altri Scarti Industriali per la Formulazione di Materiali Alleggeriti con Effetto Fertilizzante e di Materiali Polimerici per Isolamento Termico*; University of Modena and Reggio Emilia: Emilia, Italy, 2018.

24. European Coffee Federation. European Coffee Report 2017–2018. 2019. Available online: https://www.ecf-coffee.org/publications/european-coffee-report (accessed on 1 September 2020).

25. Comitato Italiano del Caffè. Esportazione Caffè 2017–2019. Available online: http://comitcaf.it/index.php/esportazione-caffe/ (accessed on 30 July 2020).

26. Allesina, G.; Pedrazzi, S.; Lovato, F.; Allegretti, F.; Tartarini, P.; Siligardi, C. Discussion of possible coffee grounds disposal chains for energy production. In Proceedings of the 23rd EUBCE (European Biomass Conference and Exhibition), Wien, Austria, 1–4 June 2015.

27. Van Zwieten, L.; Kimber, S.; Sinclair, K.; Chan, K.Y.; Downie, A. Biochar: Potential for climate change mitigation, improved yield and soil health. In *Pastures at the Cutting Edge: Proceedings of the 23rd Annual Conference of the Grassland Society of NSW, Tamworth, Australia, 21–23 July 2008*; Boschma, S.P., Serafin, L.M., Ayres, J.F., Eds.; Grassland Society of NSW Inc: Orange, Australia, 2008; pp. 30–33152. 152p.

28. Yamato, M.; Okimori, Y.; Wibowo, I.F.; Anshori, S.; Ogawa, M. Effects of the application of charred bark of Acacia mangium on the yield of maize, cowpea and peanut, and soil chemical properties in South Sumatra, Indonesia. *Soil Sci. Plant Nutr.* **2006**, *52*, 489–495. [CrossRef]

29. Glaser, B.; Lehmann, J.; Zech, W. Ameliorating physical and chemical properties of highly weathered soils in the tropics with charcoal—A review. *Biol. Fertil. Soils* **2002**, *35*, 219–230. [CrossRef]

30. Windeatt, J.H.; Ross, A.A.; Williams, P.T.; Forster, P.M.; Nahil, M.A.; Singh, S. Characteristics of biochars from crop residues: Potential for carbon sequestration and soil amendment. *J. Environ. Manag.* **2014**, *146*, 189–197. [CrossRef] [PubMed]

31. Decree, L. *Decreto Legislativo n. 75, 29 Aprile 2010: Riordino e Revisione Della Disciplina in Materia di Fertilizzanti, a Norma Dell'articolo 13 Della Legge 7 Luglio 2009 n. 88. Gazzetta Ufficiale della Repubblica Italiana 2010, 106*; European Parliament and European Council. Regulation (EC) No 2003/2003 of the European Parliament and of the Council Related to Fertilizers; European Union: Brussels, Belgium, 2003.

32. Lorenz, D.H.; Eichhorn, K.W.; Blei-Holder, H.; Klose, R.; Meier, U.; Weber, E. Phenological growth stages and BBCH—Identification keys of grapevine (*Vitis vinifera* L. ssp. vinifera). In *Growth Stages of Mono and Dicotyledonous Plants—BBCH Monograph*, 2nd ed.; Meier, U., Ed.; Federal Biological Research Centre for Agriculture and Forestry: Berlin, Germany, 2001; p. 158.

33. Cerovic, Z.G.; Masdoumier, G.; Ghozlen, N.B.; Latouche, G. A new optical leaf-clip meter for simultaneous non-destructive assessment of leaf chlorophyll and epidermal flavonoids. *Physiol. Plant* **2012**, *146*, 251–260. [CrossRef] [PubMed]

34. Ronga, D.; Pentangelo, A.; Parisi, M. Optimizing N Fertilization to Improve Yield, Technological and Nutritional Quality of Tomato Grown in High Fertility Soil Conditions. *Plants* **2020**, *9*, 575. [CrossRef]

35. Waite, H.; Whitelaw-Weckert, M.; Torley, P. Grapevine propagation: Principles and methods for the production of high-quality grapevine planting material. *N. Z. J. Crop Hortic. Sci.* **2015**, *43*, 144–161. [CrossRef]

36. Valdés-Gómez, H.; Celette, F.; García de Cortázar-Atauri, I.; Jara-Rojas, F.; Ortega-Farías, S.; Gary, C. Modelling soil water content and grapevine growth and development with STICS crop-soil model under two different water management strategies. *J. Int. Sci. Vigne Vin* **2009**, *43*, 13–28. [CrossRef]

37. Carey, D.E.; McNamara, P.J.; Zitomer, D.H. Biochar from pyrolysis of biosolids for nutrient adsorption and turfgrass cultivation. *Water Environ. Res.* **2015**, *87*, 2098–2106. [CrossRef]

38. Caradonia, F.; Ronga, D.; Catellani, M.; Giaretta Azevedo, C.V.; Terrazas, R.A.; Robertson-Albertyn, S.; Francia, E.; Bulgarelli, D. Nitrogen fertilizers shape the composition and predicted functions of the microbiota of field-grown tomato plants. *Phytobiomes J.* **2019**, *3*, 315–325. [CrossRef]

39. Prichard, T. Vineyard Irrigation Systems. In *Raisin Production Manual University of California Agricultural and Natural Resources*; Apex: Oakland, CA, USA, 2000; Volume 3393, pp. 57–63.

40. Richards, D. The grape root system. *Hortic. Rev.* **1983**, *5*, 127–168.

41. Clarke, S.J.; Lamont, K.J.; Pan, H.Y.; Barry, L.A.; Hall, A.; Rogiers, S.Y. Spring root-zone temperature regulates root growth, nutrient uptake and shoot growth dynamics in grapevines. *Aust. J. Grape Wine Res.* **2015**, *21*, 479–489. [CrossRef]

42. Rogiers, S.Y.; Clarke, S.J.; Schmidtke, L.M. Elevated root-zone temperature hastens vegetative and reproductive development in Shiraz grapevines. *Aust. J. Grape Wine Res.* **2014**, *20*, 123–133. [CrossRef]

43. Jacobs, D.F.; Timmer, V.R. Fertilizer-induced changes in rhizosphere electrical conductivity: Relation to forest tree seedling root system growth and function. *New For.* **2005**, *30*, 147–166. [CrossRef]

44. Abad, M.; Noguera, P.; Bures, S. National inventory of organic wastes for use as growing media for ornamental potted plant production: Case study in Spain. *Bioresour. Technol.* **2001**, *77*, 197–200. [CrossRef]

45. Ronga, D.; Pane, C.; Zaccardelli, M.; Pecchioni, N. Use of spent coffee ground compost in peat-based growing media for the production of basil and tomato potting plants. *Commun. Soil Sci. Plant* **2016**, *47*, 356–368. [CrossRef]

46. Schmidt, H.P.; Kammann, C.; Niggli, C.; Evangelou, M.W.; Mackie, K.A.; Abiven, S. Biochar and biochar-compost as soil amendments to a vineyard soil: Influences on plant growth, nutrient uptake, plant health and grape quality. *Agric. Ecosyst. Environ.* **2014**, *191*, 117–123. [CrossRef]

47. Vaccari, F.P.; Maienza, A.; Miglietta, F.; Baronti, S.; Di Lonardo, S.; Guagnoni, L.; Lagomarsino, A.; Pozzi, A.; Pusceddu, E.; Ranieri, R.; et al. Biochar stimulates plant growth but not fruit yield of processing tomato in a fertile soil. *Agric. Ecosyst. Environ.* **2015**, *207*, 163–170. [CrossRef]

48. Sigua, G.C.; Novak, J.M.; Watts, D.W.; Johnson, M.G.; Spokas, K. Efficacies of designer biochars in improving biomass and nutrient uptake of winter wheat grown in a hard setting subsoil layer. *Chemosphere* **2016**, *142*, 176–183. [CrossRef]

49. Chalker-Scott, L. Environmental significance of anthocyanins in plant stress responses. *Photochem. Photobiol.* **1999**, *70*, 1–9. [CrossRef]

50. Zhao, L.; Cao, X.; Zheng, W.; Scott, J.W.; Sharma, B.K.; Chen, X. Copyrolysis of biomass with phosphate fertilizers to improve biochar carbon retention, slow nutrient release, and stabilize heavy metals in soil. *ACS Sustain. Chem. Eng.* **2016**, *4*, 1630–1636. [CrossRef]

51. Gwenzi, W.; Nyambishi, T.J.; Chaukura, N.; Mapope, N. Synthesis and nutrient release patterns of a biochar-based N–P–K slow-release fertilizer. *Int. J. Environ. Sci. Technol.* **2018**, *15*, 405–414. [CrossRef]

52. Setti, L.; Francia, E.; Pulvirenti, A.; Gigliano, S.; Zaccardelli, M.; Pane, C.; Caradonia, F.; Bortolini, S.; Maistrello, L.; Ronga, D. Use of black soldier fly (*Hermetia illucens* (L.), Diptera: Stratiomyidae) larvae processing residue in peat-based growing media. *Waste Manag.* **2019**, *95*, 278–288. [CrossRef] [PubMed]

53. Raviv, M.; Oka, Y.; Katan, J.; Hadar, Y.; Yogev, A.; Medina, S.; Krasnovskya, A.; Ziadna, H. High-nitrogen compost as a medium for organic container-grown crops. *Bioresour. Technol.* **2005**, *96*, 419–427. [CrossRef] [PubMed]

54. Amendola, C.; Montagnoli, A.; Terzaghi, M.; Trupiano, D.; Oliva, F.; Baronti, S.; Miglietta, F.; Chiatante, D.; Scippa, G.S. Short-term effects of biochar on grapevine fine root dynamics and arbuscular mychorrhizae production. *Agric. Ecosyst. Environ.* **2017**, *239*, 236–245. [CrossRef]

55. Gaiotti, F.; Marcuzzo, P.; Belfiore, N.; Lovat, L.; Fornasier, F.; Tomasi, D. Influence of compost addition on soil properties, root growth and vine performances of *Vitis vinifera cv* Cabernet sauvignon. *Sci. Hortic.* **2017**, *225*, 88–95. [CrossRef]

56. Bernal-Vicente, A.; Ros, M.; Tittarelli, F.; Intrigliolo, F.; Pascual, J.A. *Citrus* compost and its water extract for cultivation of melon plants in greenhouse nurseries. Evaluation of nutriactive and biocontrol effects. *Bioresour. Technol.* **2008**, *99*, 8722–8728. [CrossRef]

57. Ronga, D.; Caradonia, F.; Setti, L.; Hagassou, D.; Giaretta Azevedo, C.V.; Milc, J.; Pedrazzi, S.; Allesina, G.; Arru, L.; Francia, E. Effects of innovative biofertilizers on yield of processing tomato cultivated in organic cropping systems in northern Italy. *Acta Hortic.* **2019**, *1233*, 129–136. [CrossRef]

58. Herrera, F.; Castillo, J.E.; Chica, A.F.; López Bellido, L. Use of municipal solid waste compost (MSWC) as a growing medium in the nursery production of tomato plants. *Bioresour. Technol.* **2008**, *99*, 287–296. [CrossRef]

59. Ericsson, T.; Rytter, L.; Vapaavuori, E. Physiology of carbon allocation in trees. *Biomass Bioenergy* **1996**, *11*, 115–127. [CrossRef]

60. Ronga, D.; Setti, L.; Salvarani, C.; De Leo, R.; Bedin, E.; Pulvirenti, A.; Mil, J.; Francia, E. Effects of solid and liquid digestate for hydroponic baby leaf lettuce (*Lactuca sativa* L.) cultivation. *Sci. Hortic.* **2019**, *244*, 172–181. [CrossRef]

61. Leghari, S.J.; Wahocho, N.A.; Laghari, G.M.; HafeezLaghari, A.; MustafaBhabhan, G.; HussainTalpur, K.; Lashari, A.A. Role of nitrogen for plant growth and development: A review. *Adv. Environ. Biol.* **2016**, *10*, 209–219.

62. Araujo, F.; Williams, L.E.; Matthews, M.A. A comparative study of young 'Thompson Seedless' grapevines (*Vitis vinifera* L) under drip and furrow irrigation. II. Growth, water use efficiency and nitrogen partitioning. *Sci. Hortic.* **1995**, *60*, 251–265. [CrossRef]

63. Graber, E.R.; Tsechansky, L.; Mayzlish-Gati, E.; Shema, R.; Koltai, H. A humic substances product extracted from biochar reduces Arabidopsis root hair density and length under P-sufficient and P-starvation conditions. *Plant Soil* **2015**, *395*, 21–30. [CrossRef]

64. Jindo, K.; Martim, S.A.; Navarro, E.C.; Pérez-Alfocea, F.; Hernandez, T.; Garcia, C.; Aguiar, N.O.; Canellas, L.P. Root growth promotion by humic acids from composted and non-composted urban organic wastes. *Plant Soil* **2012**, *353*, 209–220. [CrossRef]

horticulturae

Article

Nitrogen Effect on Growth-Related Parameters and Evaluation of *Portulaca oleracea* as a Phytoremediation Species in a Cr(VI)-Spiked Soil

Georgios Thalassinos, Elina Nastou, Spyridon A. Petropoulos * and Vasileios Antoniadis *

Department of Agriculture, Crop Production and Rural Environment, School of Agricultural Sciences, University of Thessaly, 38446 Volos, Greece; thalassinosgeorgios@hotmail.gr (G.T.); elina_90@windowslive.com (E.N.)
* Correspondence: spetropoulos@uth.gr (S.A.P.); antoniadis@uth.gr (V.A.); Tel.: +30-242-109-3196 (S.A.P.); +30-242-109-3241 (V.A.)

Citation: Thalassinos, G.; Nastou, E.; Petropoulos, S.A.; Antoniadis, V. Nitrogen Effect on Growth-Related Parameters and Evaluation of *Portulaca oleracea* as a Phytoremediation Species in a Cr(VI)-Spiked Soil. *Horticulturae* 2021, 7, 192. https://doi.org/10.3390/ horticulturae7070192

Academic Editor: Rui Manuel Almeida Machado

Received: 8 June 2021
Accepted: 12 July 2021
Published: 14 July 2021

Publisher's Note: MDPI stays neutral with regard to jurisdictional claims in published maps and institutional affiliations.

Abstract: In a pot experiment, we assessed the potential of purslane (*Portulaca oleracea*) as a phytoremediation species in Cr(VI)-contaminated soils. We focused on the evaluation of phytotoxic Cr(VI) effects at concentrations reaching 150 mg Cr(VI) kg^{-1} and the possible stress amelioration effect of nitrogen on Cr(VI)-stressed plants. Treatments were T-0 (control), T-1 (25 mg Cr(VI) kg^{-1}), T-2 = 50 mg kg^{-1}, T-3 = 100 mg kg^{-1}, and T-4 = 150 mg kg^{-1}. We measured Cr(VI) concentration in aerial and root tissues, a series of parameters related to photosynthesis and plant growth, phosphorus aerial plant tissue content, and we also calculated indices (ratios) related to leaf growth and above ground tissue water content. Cr(VI) almost exclusively was found in root tissues; all physiological and growth parameters studied were severely affected and plants selectively accumulated phosphorus in aerial plant tissues with increasing Cr(VI) soil concentrations. On the other hand, N amendment resulted in improved plant features in some of the measured parameters: chlorophyll index was improved with added N at T-2, plant height was significantly higher at T-0, T-1, and T-2, and aerial dry weight and leaf area was higher at T-0; these effects indicate that added N did increase P. oleracea potential to ameliorate Cr(VI) toxic effects. We conclude that purslane showed a potential as a possible species to be successfully introduced to Cr(VI)-laden soils, but more research is certainly necessary.

Keywords: hexavalent Cr; photosynthesis; phosphorus uptake; Cr(VI) tissue; leaf characteristics; purslane; soil contamination; heavy metals

1. Introduction

Metal ions can be introduced to surface soils by natural or anthropogenic processes and their environmental impact is greatly affected by their mineralogical and geochemical form [1]. Cr is mainly found in two valence states, namely +3 (chromite (Cr(III)) and +6 (chromate Cr(VI)). However, in natural soil conditions trivalent chromium Cr(III) is the predominant state [2]. Hexavalent Cr compounds are found in wastes of numerous industrial activities (i.e., chromic acid and Cr-pigment production, leather tanning, cement production, metal plating, and stainless-steel production), and its anionic form results in increased possibility of Cr(VI) pollution dispersal [2–4].

Cr(III) is an essential element for some redox reactions that serve fundamental cellular functions relevant to sugar, protein and lipid metabolism in humans (recommended adult intake of 50 to 200 µg/d); however it is not an essential element for plants [5–7]. Hexavalent chromium (Cr(VI)) is of much higher toxicity (10 to 100 times) compared to Cr(III) for both acute and chronic exposure, posing serious health hazards for humans. Hexavalent Cr has been identified as one of the seventeen chemicals threatening human health and is classified as a human carcinogen causing a variety of cancer diseases in humans that result in increased overall mortality rates. Cr(VI) and Cr(III) in soil are in dynamic equilibrium and the concentration of each form depends primarily on soil characteristics and redox conditions [5,7–9].

The particularly high toxicity of Cr(VI) in prokaryotes and eukaryotes is attributed to the high bioavailability of Cr(VI) oxyanions found in the soil environment (i.e., chromate (CrO_4^{2-}), hydrogen-chromate ($HCrO_4^-$), and dichromate ($Cr_2O_7^{2-}$). In common soil pH values, soil constituents bear mainly negative charges resulting in limited anion binding capacity [4,5,10].

Inside cells, the metabolic reduction of Cr(VI) through enzymatic and non-enzymatic processes leads to the formation of Cr(III) and in the parallel production of reactive oxygen species ROS, resulting in severe oxidative damage to plant cells. Cr(III) remains inside plant cells because of its low membrane permeability, forming stable complexes with proteins and nucleic acids resulting in the inhibition of DNA replication and RNA transcription [8,11–14].

In plants, translocation and accumulation of Cr largely depends on Cr speciation, plant species-specific stress alleviating mechanisms, concentration of Cr and Cr availability in the growth medium [11,15,16]. Cr(VI) enter plant cells via the cell membrane generic anion channels, due to structural similarities of chromate to sulfate and phosphate anions [12,13]. Contrary to that, Cr(III) ions can cross cell membrane at a much lower rate via simple diffusion or pinocytosis [17,18]. Several studies indicated that Cr primarily accumulated in the plant roots [19]. Cr(VI) is readily sequestered in root vacuoles and is poorly transported to aerial biomass in an effort of the plant to address Cr(VI) toxicity, thus avoiding exposure of important aerial organs for its physiological functions to elevated Cr(VI) [2,11].

It has been proposed that mechanisms developed from plant species tolerant to abiotic stresses, contribute to heavy metal tolerance. Cross-tolerance mechanisms between abiotic stresses and heavy metal tolerance mechanisms have been reported for various plant species [20]. Halophyte plant species have developed a series of mechanisms that confer tolerance to many metal ions, in concentrations prohibitive to the growth of most plant species [21,22]. The mechanisms implemented from halophytes include vacuole accumulation of metal ions, exclusion of metal ions from entering root cells and excretion of metal ions through the salt glands [23]. Furthermore, in halophytes various organic compounds may be accumulated, for cells to maintain their structure and protect the function of enzymatic mechanisms due to salt stress [24]. In halophytic plant species, increased concentrations of proline, total soluble sugars, and amino acids (such as leucine, isoleucine, valine, glutamine, glutamate, tyrosine, threonine, arginine, phenylalanine, and tryptophan) have been reported as a response to elevated toxic metal concentrations [25,26]. The above mechanisms result in the alleviation of heavy metal stress, rendering halophytes as potential candidates for phytoextraction and phytostabilization, as well as for saline agriculture [20,23,27,28].

Overall, high levels of Cr in plant tissues result in reduced plant height, root length, chlorophyll and pigment content in leaves, reduced photosynthetic rate, damaged root tissues, ultrastructural modifications of cell membranes, mineral nutrient imbalance and reduced enzymatic activity [6,14,29]. Chromium can limit the absorption of elements essential for plant growth such as N, P, K, Fe, Mg, Mn, Mo, Zn, Cu, Ca, and B [11,30,31]. Moreover, Cr(VI) has a negative effect on enzymes relevant to nitrogen metabolism, decreasing the activity of nitrate and nitrite reductase, glutamine synthetase, urease, and glutamate dehydrogenase [2,11,32].

Nutrient supply of micro- and macronutrients is an essential factor influencing plant growth that helps to alleviate the negative effects of biotic and abiotic stressors on plants. However, under field conditions only a limited number of nutrients have been found to alleviate biotic and abiotic stresses, among which, sulfate [33]. On the other hand, it is known that species exposed to a certain environmental restrictive agent, may become less effective in addressing other additional stresses [34,35]. In that sense, it should be expected that plants growing in unfertilized soils could be severely affected upon exposure to Cr(VI) contamination. However, to the best of our knowledge, this has never been tested before in real soil conditions. Thus, it could be the case that well-fertilized plants address Cr(VI) exposure in a way that their developmental (i.e., root and shoot weight, and aerial part height), as well as physiological features (i.e., photosynthetic rate and chlorophyll content) are less severely affected compared to non-fertilized plants; however, due to a lack of evidence from the literature, this potentially beneficial effect of fertilization has to be elucidated [36].

Purslane (*Portulaca oleracea*) is a halophytic annual plant species, tolerant to several abiotic stresses [22]. This species adaptability is largely attributed to great morpho-cyto-physiologic variability that greatly contributes to the rapid growth and propagation under harsh environmental conditions. Key factors that contribute to purslane adaptability involve the production of secondary metabolites and the species ability to switch from C4 to CAM photosynthesis (carbon fixation-photosynthetic mechanism) under drought stress [21,37]. Salinity, drought and metal stress induce common physiological responses from plants [38]. Tolerant plants to other abiotic stresses can be possible candidates to be tested for phytoremediation purposes. To the best of our knowledge, despite the literature reports regarding the accumulation and toxic effects of chromium on purslane plants [36,39–41], there is a void in the literature concerning the ability of the species to grow under elevated Cr(VI) soil conditions, especially when N-added soils and compared to soil with non N addition.

Furthermore, it may be the case that nitrogen applied to Cr(VI)-stressed purslane could result in higher plant aerial biomass and thus in higher overall removal of Cr(VI) from soil. If this is the case, purslane could act faster as a phytoremediation species for the restoration of a Cr(VI)-laden area, when considering that metal uptake is affected both by plant tissue Cr(VI) content and the aerial biomass. However, this beneficial effect is also not elucidated by current literature evidence. Hence, the aims of this work were to study the developmental and physiological features of purslane, as well as its Cr(VI) content and possible toxicity symptoms resulting from Cr(VI) exposure in a soil well-fertilized with N and compare these effects with those in purslane plants grown in a non-fertilized soil. This study targeted specifically the evaluation of purslane as a potential phytoremediation species towards Cr(VI)-contaminated soils. To the best of our knowledge, although there are some works that have investigated purslane as a phytoremediation species towards Cr(VI) (e.g., [36]), there is no investigation in relation to the effects of added N on the phytoremediation capacity of the species.

2. Materials and Methods

A 10-treatment (2 levels of nitrogen $\times$ 5 levels of Cr(VI)) $\times$ 10 replicates) experiment was established. Overall, we had 100 replicates (each in 2-L pots) and for each replicate, a mixture of 1000 g of soil and 800 mL perlite was prepared. Soil was obtained from a field in the agricultural region between Volos and Larisa (39.394925 N, 22.756285 E), an area not affected from any known source of pollution. Soil spiking was performed using Cr(VI) solution of 10,000 ppm Cr(VI), by dissolving 19.22 g of CrO_3 in 1000 mL distilled water. Spiking solution was applied to the soil resulting in 5 Cr(VI) treatments (T-0: control; T-1: 25 mg Cr(VI) kg^{-1}, with 2.5 mL spiking solution per pot; T-2: 50 mg kg^{-1}, with 5 mL solution; T-3: 100 mg kg^{-1}, 10 mL solution; and T-4: 150 mg kg^{-1}, 15 mL solution.

For each Cr(VI) treatment, half of the replicates (10 out of 20) were amended in rates equivalent to 200 kg of nitrogen per hectare or 100 mg N per kg soil as NH_4NO_3 salt (thereafter named N-1 treatments and the non-added-N treatments are named N-0). The spiked soil treatments along with the un-amended control were placed in 2-L plastic pots, watered to their holding capacity and the spiked soil was left to equilibrate for 20 days. During the equilibration period, soil was thoroughly mixed three times per week and water was added as needed to keep soil to its water holding capacity.

At the end of the equilibration period, four samples per Cr(VI) treatment were obtained from the pots, air dried, passed through a 2-mm sieve in order to determine the initial (Day 0) hexavalent chromium (Cr(VI)) soil concentration.

2.1. Plant Establishment, Measurements and Soil and Plant Analysis

Cr(VI) was extracted from soil samples using 0.01 M KH_2PO_4 solution, color was developed by the diphenyl carbazide method and absorption values were determined using a Biochrom Libra S11 spectrophotometer at 540 nm [42]. For each treatment the initial hexavalent chromium (Cr(VI)) concentrations were estimated (T-0: 0.0 mg Cr(VI) kg^{-1} (control), i.e., un-amended soil; T-1: 20.65 mg Cr(VI) kg^{-1}; T-2: 49.92 mg Cr(VI) kg^{-1}; T-3: 106.43 mg Cr(VI) kg^{-1}; and T-4: 148.62 mg Cr(VI) kg^{-1}). Furthermore, three randomly acquired samples were analyzed for soil physiochemical parameters (pH = 7.8, ECe = 850 μS cm^{-1}, $CaCO_3$ = 10.4%, soil of loam texture) according to commonly used laboratory protocols [43].

Plants were grown in an unheated greenhouse. On Day 0, *P. oleracea* plants, already sown 25 days before Day 0 in peat-filled seedling trays, were transplanted in pots (one plant per pot). Transplantation took place when plants reached a height of 12 cm.

During the growth period, to compensate for any light and temperature differences in the greenhouse, plant positions were exchanged regularly, and water was applied to the plants according at regular intervals in amounts that depended on weather conditions (50–250 mL per pot). One month prior to harvest date, we measured plant height in cm, photosynthetic rate (μmol CO_2 m^{-2} s^{-1}) at a constant light intensity (250 μmol cm^{-2} s^{-1}) using the LI-Cor LI-6400XT Portable Photosynthesis System (LI-Cor, Lincoln, NE, USA)) and chlorophyll content (SPAD index) was measured using the OPTI-SCIENCES CCM-200 plus chlorophyll content meter (Opti-sciences, Hudson, NH, USA).

Plants were grown in the Cr(VI) spiked soil for 50 days—from 14 October 2019 (establishment of seedlings in the pots) to 4 December 2019, when plants were harvested. On harvest day we measured the weight of stems, the weight of leaves and the leaf area per plant. Then, aerial plant tissues were washed with deionized water and root tissues were meticulously washed so that no soil particles remained attached and further rinsed with deionized water.

Aerial and root plant tissues were dried in an air-forced drying oven at 70 °C for 96 h. Both aerial and root tissues were weighted and pulverized. Then, 1.00 g samples of plant tissue were dry-ashed at 500 °C for 4 h and ash was extracted using 10 mL of 20% HCl. Plant tissue K, P and Cr(VI) concentrations were estimated according to established laboratory protocols—dry ashing at 500 °C for 5 h, and then ash extraction with 20 mL of 20% HCl [44]. Due to the lack of sufficient plant tissue mass, especially in the high Cr(VI) treatments, out of the 10 replicates initially sown, 5 replicates for extraction and measurement were formed by combining tissues from every two pots. Furthermore, out of the primary data, we calculated a secondary index, i.e., tolerance index (TI), equal to dry aerial biomass in contaminated soil over that in control. Because of the fact that we had effectively two controls, typical to a two-factor experiment like ours, i.e., (a) no Cr(VI) with no N, and (b) no Cr(VI) with added N, we calculated TI as two independent factors, one for soils without N and one for soils with N.

2.2. Quality Assurance and Statistical Analyses

For data quality control purposes in-house plant and soil reference materials were used and recovery rates were within the range of 95% to 105% of the certified value. To rule out any possibility of cross-contamination, for every extraction batch blank samples were also measured. For Cr(VI) calibration curves, Merck standard solutions were used (Merck, Burlington, MA, USA).

Statistical analysis of the data was performed using IBM SPSS Statistics 25 and Excel 2019 software. One-way ANOVA and Duncan's multiple range tests were used to identify statistically significant differences between treatments and two-way ANOVA and Duncan's multiple range tests at $p = 0.05$ were used to monitor the effect that Cr and nitrogen had on the different parameters studied.

3. Results

3.1. Cr(VI) Concentration in Plant Tissues

Increasing soil Cr(VI), increased aerial tissue Cr(VI) content ($p < 0.001$), reaching 4.13 mg Cr(VI) kg^{-1} dry matter (T-4 (Cr(VI)) treatment with no added N), while nitrogen addition had no significant effect in the aerial tissue Cr(VI) content ($p = 0.915$) (Figure 1a). Furthermore, increasing Cr(VI) soil concentration resulted in significant increase ($p < 0.001$) in root tissue Cr(VI) concentration and nitrogen addition had no significant effect on root tissue Cr(VI) content ($p = 0.109$). In root tissues, Cr(VI) levels were orders of magnitude higher compared to aerial tissues, reaching 339 mg kg^{-1} dry matter at the highest Cr(VI) soil concentration with no added N, while at T-4 Cr(VI) with added N root Cr(VI) concentrations reached 596 mg kg^{-1} ($p < 0.001$). This finding indicates that in the highest tested soil Cr(VI) concentrations, N amendment resulted in increased root Cr(VI) concentrations (Figure 1b), significantly higher than the non-added-N treatment, whereas aerial contents of Cr(VI) remained low without being affected by N addition.

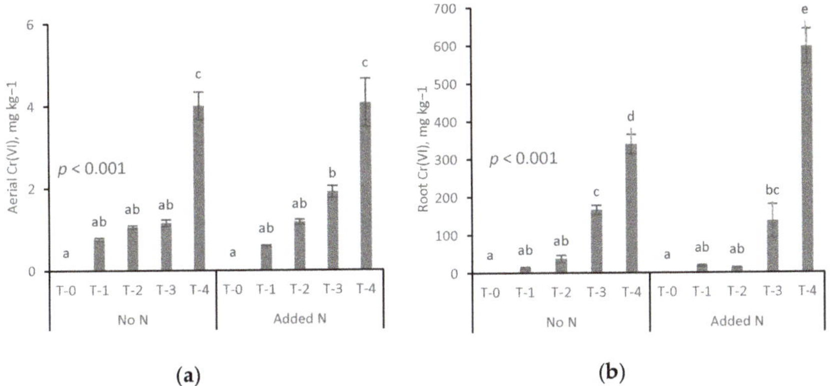

Figure 1. Cr(VI) concentration in plant tissues: (a) Aerial tissue Cr(VI) concentrations (mg Cr(VI) kg^{-1} dry matter); (b) Root tissue Cr(VI) concentrations (mg Cr(VI) kg^{-1} dry matter). Different letters denote statistically significant differences at $p < 0.05$. Error bars represent standard error (T-1 = 25, T-2 = 50, T-3 = 100, T-4 = 150 mg Cr(VI) kg^{-1} soil).

3.2. Effects of Cr(VI) on Parameters Relative to Photosynthesis and Plant Growth

3.2.1. Chlorophyll Content Index and Photosynthetic Rate

As a result of Cr(VI) exposure, the purslane developmental and physiological parameters studied were significantly affected. Chlorophyll content (SPAD index) was found to gradually decrease ($p < 0.001$) from 12.35 ((T-0)-no added N) to 4.82 (T-4-no added N). Added nitrogen resulted in significantly higher chlorophyll content (Figure 2a). Similarly, with increasing Cr(VI) soil concentrations, reduced photosynthetic rate values were noticed ($p < 0.001$) and nitrogen addition resulted in increased ($p = 0.020$) photosynthetic rate values for every Cr(VI) level (Figure 2b).

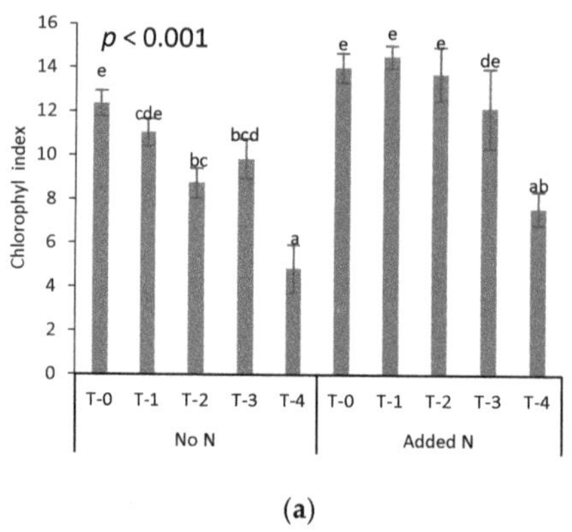

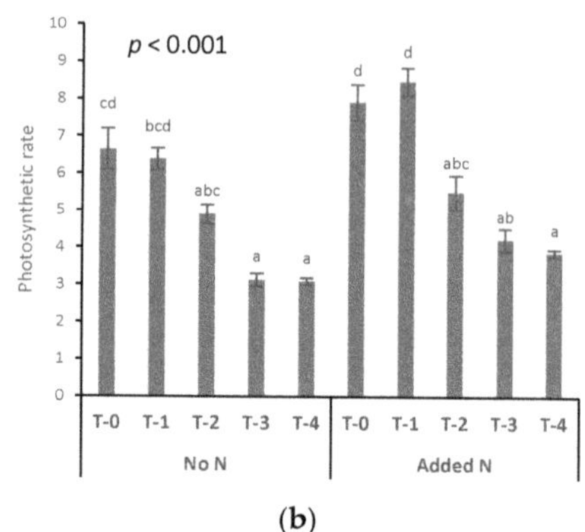

(a)   (b)

Figure 2. (**a**) Chlorophyll index; (**b**) Photosynthetic rate (μmol CO_2 m^{-2} s^{-1}). Different letters denote statistically significant differences at $p < 0.05$. Error bars represent standard error (T-1 = 25, T-2 = 50, T-3 = 100, T-4 = 150 mg Cr(VI) kg^{-1} soil).

3.2.2. Plant Height and Aerial Fresh Weight

Plant height and aerial fresh weight, as expected, followed the same trend observed for photosynthetic rate and chlorophyll content. Increasing Cr(VI) concentrations resulted in lower height values ($p < 0.001$) and nitrogen amendment had a positive effect in plant height in every Cr(VI) level tested ($p = 0.038$). For aerial fresh weight the trend was similar, with increasing Cr(VI) concentrations exerting negative effects on the measured values ($p < 0.001$) and in this case nitrogen addition had also a positive effect ($p < 0.001$) (Table 1).

Table 1. Plant height and aerial fresh weight at harvest date.

Treatments		Plant Height	Aerial Fresh Weight
		(cm)	(g pot^{-1})
No N	T-0	29.9 ± 1.09 [f]	13.4 ± 2.18
	T-1	18.9 ± 0.57 [d]	5.9 ± 0.58
	T-2	16.3 ± 0.91 [c]	4.4 ± 0.33
	T-3	15.0 ± 0.41 [bc]	4.5 ± 0.35
	T-4	11.2 ± 0.46 [a]	2.3 ± 0.31
Added N	T-0	31.2 ± 1.39 [d]	29.5 ± 3.21
	T-1	22.0 ± 0.67 [bc]	8.2 ± 0.52
	T-2	15.7 ± 0.48 [ab]	5.7 ± 0.30
	T-3	15.7 ± 0.99 [ab]	4.2 ± 0.38
	T-4	13.5 ± 0.46 [a]	3.5 ± 0.40
Treatment effect		$p < 0.001$	$p = 0.298$
Cr effect		$p < 0.001$	$p < 0.001$
Nitrogen effect		$p = 0.038$	$p < 0.001$

Mean ± S. E. Different letters denote significant ($p < 0.05$) difference between means in columns according to Duncan's multiple range test (T-1 = 25, T-2 = 50, T-3 = 100, T-4 = 150 mg Cr(VI) kg^{-1} soil).

3.2.3. Aerial Dry Weight and Root Dry Weight and Leaf Area per Plant

Cr(VI) increasing concentrations negatively affected aerial dry matter production ($p < 0.001$), and nitrogen addition resulted in higher values for every Cr(VI) level studied ($p < 0.001$). More specifically, in treatments where no nitrogen was added, values decreased gradually from 0.99 g (T-0) to 0.23 g (T-4) and in treatments where nitrogen was added, aerial dry weight values were significantly higher, reaching 2.73 g in the control treatment (T-0) and gradually decreased to 0.36 g for the highest soil Cr(VI) concentration used in the experiment (T-4). On the other hand, root dry weight showed an increasing trend with increasing Cr(VI) concentrations even from the lower Cr(VI) level (T-1) (0.39 g per pot), despite the fact that differences failed to escalate in the higher Cr(VI) levels, resulting in marginally higher than 0.05 significance ($p = 0.053$); nitrogen addition resulted in higher root dry weight values ($p = 0.021$) (Table 2). Considering the results presented in Figure 1 and Table 2, the addition of N in plants treated with the highest Cr(VI) concentration may result in the removal of significantly higher amounts of Cr(VI) from contaminated soils. In particular, despite the similar contents of Cr(VI) in the aerial parts the increase in dry weight per pot (from 0.23 g pot^{-1} to 0.36 g pot^{-1} in no N and added N plants, respectively) indicates an increase in the removal of Cr(VI) from soils by 56.5%. On the other hand, in the case of roots we noticed an increase in both root dry weight per pot (from 0.5 g pot^{-1} to 0.96 g pot^{-1} in no N and added N plants, respectively) and Cr(VI) content in dried tissues (339 mg kg^{-1} and 596 mg kg^{-1} in no N and added N plants, respectively), meaning a cumulative increase of 237.6% in Cr(VI) removal from soils.

Table 2. Fresh and dry weight of aerial tissues, and dry weight of root tissues.

Treatments		Aerial Dry Weight (g pot^{-1})	Root Dry Weight (g pot^{-1})
No N	T-0	0.99 ± 0.167 [b]	0.30 ± 0.089 [a]
	T-1	0.53 ± 0.009 [ab]	0.39 ± 0.053 [ab]
	T-2	0.42 ± 0.015 [a]	0.56 ± 0.084 [abc]
	T-3	0.48 ± 0.049 [ab]	0.83 ± 0.093 [abc]
	T-4	0.23 ± 0.015 [a]	0.50 ± 0.127 [abc]
Added N	T-0	2.73 ± 0.339 [c]	0.38 ± 0.056 [ab]
	T-1	0.76 ± 0.107 [ab]	0.76 ± 0.169 [abc]
	T-2	0.50 ± 0.019 [ab]	1.02 ± 0.126 [c]
	T-3	0.45 ± 0.051 [a]	0.79 ± 0.095 [abc]
	T-4	0.36 ± 0.052 [a]	0.96 ± 0.146 [bc]
Treatment effect		$p < 0.001$	$p = 0.045$
Cr effect		$p < 0.001$	$p = 0.053$
Nitrogen effect		$p < 0.001$	$p = 0.021$

Mean $\pm$ S. E. Different letters denote significant ($p < 0.05$) difference between means in columns according to Duncan's multiple range test (T-1 = 25, T-2 = 50, T-3 = 100, T-4 = 150 mg Cr(VI) kg^{-1} soil).

3.2.4. Leaf Area per Plant, Leaf Weight/Total Aerial Weight Ratio and Aerial Tissue Dry Matter Content (Aerial Dry Weight/Aerial Fresh Weight Ratio)

Dry to fresh aerial tissue weight ratio increased significantly ($p = 0.002$) even from the lowest soil Cr(VI) (T-1) concentration (Table 3). Leaf area was significantly reduced ($p < 0.001$) even from the lowest level of Cr(VI) soil concentration (T-1), while nitrogen addition had a positive effect, increasing leaf area ($p < 0.001$) (Table 3). Apart from leaf area, leaf weight-to-total aerial weight ratio followed the same pattern as the leaf area, with increasing Cr(VI) concentrations resulting in lower ($p < 0.001$) ratio values and N addition resulting in higher ratio values ($p = 0.007$) (Table 3). Leaf weight/total aerial tissue weight ratio decreased gradually with increasing Cr(VI) concentrations ($p < 0.001$) and nitrogen addition resulted in significantly higher ratio values (Table 3).

Table 3. Leaf area, leaf/total aerial weight ratio and fresh weight to dry weight ratio of aerial tissues.

Treatments		Leaf Area	Leaf/Aerial Weight Ratio	Aerial Tissue (Dry Weight/Fresh Weight Ratio)
		(cm^2 Plant^{-1})	(g g^{-1})	(g g^{-1})
No N	T-0	114.5 ± 18.8 [b]	0.33 ± 0.020	0.0723± 0.0032 [a]
	T-1	59.6 ± 2.5 [ab]	0.28 ± 0.021	0.0936 ± 0.0015 [bc]
	T-2	43.3 ±2.8 [a]	0.27 ± 0.028	0.0884 ± 0.0041 [bc]
	T-3	48.3 ±4.1 [a]	0.30 ± 0.022	0.0934 ± 0.0042 [bc]
	T-4	41.0 ± 3.7 [a]	0.26 ± 0.025	0.0953 ± 0.0047 [bc]
Added N	T-0	381.5 ± 19.0 [c]	0.50 ± 0.070	0.0790 ± 0.0026 [ab]
	T-1	120.9 ± 4.1 [b]	0.31 ± 0.012	0.0875 ± 0.0009 [bc]
	T-2	98.3 ± 5.9 [ab]	0.33 ± 0.018	0.0927 ± 0.004 [bc]
	T-3	68.4 ± 2.4 [ab]	0.31 ± 0.027	0.0996 ± 0.004 [c]
	T-4	49.9 ± 1.9 [a]	0.26 ± 0.024	0.0998 ± 0.008 [c]
Treatment effect		$p < 0.001$	$p = 0.358$	$p = 0.015$
Cr effect		$p < 0.001$	$p < 0.001$	$p = 0.002$
Nitrogen effect		$p < 0.001$	$p = 0.007$	$p = 0.324$

Mean ± S. E. Different letters denote significant ($p < 0.05$) difference between means in columns according to Duncan's multiple range test.

3.2.5. Tolerance Index

Tolerance index values (aerial biomass in contaminated soil/aerial biomass in control), complemented the results of physiological and growth parameters. Cr(VI) increasing soil concentrations resulted in lower tolerance index values ($p < 0.001$) and nitrogen addition had a similar effect ($p < 0.001$). It seems that nitrogen addition resulted in higher plant growth potential in the control treatment, that was abruptly limited from the toxic effect of Cr(VI), even from the lower level of Cr(VI) applied to the soil (Figure 3).

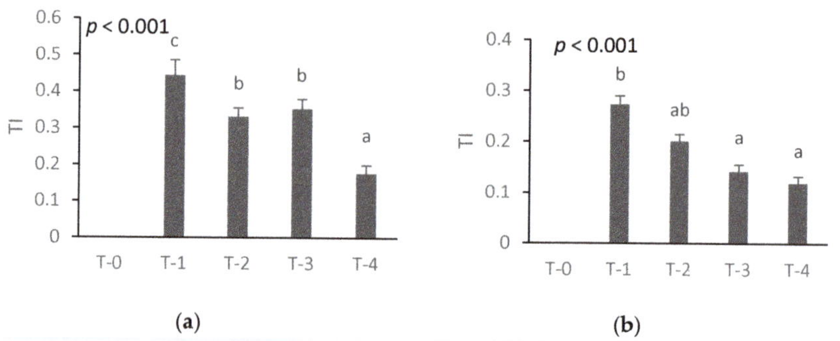

Figure 3. Tolerance index (TI) values for plants non-treated (**a**) or treated with nitrogen (**b**). Different letters denote statistically significant differences at $p < 0.05$. Error bars represent standard error (T-1 = 25, T-2 = 50, T-3 = 100, T-4 = 150 mg Cr(VI) kg^{-1} soil).

3.3. Phosphorus in Plant Tissues

In aerial plant tissues phosphorus content was measured and Duncan post hoc test results indicated that aerial tissue P concentrations increased ($p < 0.001$) with rising Cr(VI) soil concentrations, while nitrogen addition resulted in lower aerial tissue P concentrations when compared to the no N treated plants ($p = 0.034$) (Table 4). However, it must be noted that when the N effect was compared between same Cr(VI) additions, no differences were evident. For potassium content, despite the fact that the effect of Cr(VI) and N addition were significant ($p < 0.001$ and $p = 0.002$ respectively), the trend was not clear and further data are required to reach conclusive results (Table 4).

Table 4. Phosphorus and potassium content (g kg^{-1} DW) in the aerial part of purslane (*Portulaca oleracea*) aerial tissues (n = 5).

Treatments		Phosphorus Content	Potassium Content
	T-0	3.41 ab	39.50 f
	T-1	4.10 abc	28.75 abc
No N	T-2	5.10 cd	31.62 bcde
	T-3	4.72 bcd	34.65 cdef
	T-4	6.10 d	37.63 ef
	T-0	3.99 abc	35.50 def
	T-1	2.95 a	24.04 a
Added N	T-2	4.16 abc	25.92 ab
	T-3	4.40 abc	36.07 ef
	T-4	4.75 bcd	29.17 abcd
Treatment effect		$p < 0.001$	$p < 0.001$
Cr effect		$p < 0.001$	$p < 0.001$
Nitrogen effect		$p = 0.034$	$p = 0.002$

Mean $\pm$ S. E. Different superscripts denote significant ($p < 0.05$) difference between means in columns according to Duncan's multiple range test. (T-1 = 25, T-2 = 50, T-3 = 100, T-4 = 150 mg Cr(VI) kg^{-1} soil).

4. Discussion

Cr(VI) levels in root tissues were orders of magnitude higher compared to Cr(VI) concentrations found in aerial tissues, especially when plants were treated with the highest Cr(VI) concentration and fertilized with nitrogen. Increased Cr(VI) concentrations in root tissues have been noticed for a series of plant species, where plants limit the translocation of potentially toxic elements to the aerial plant tissues [2,10,11,15,45]. The physiological parameters studied (chlorophyll content and photosynthetic rate) were significantly affected with increasing Cr(VI) concentrations, while nitrogen amendment had a positive effect. Reduced values of total chlorophyll content, photosynthetic rate, and impediment of plant growth due to Cr(VI) stress have been documented for several plant species [2,8,20,26,45,46].

Values of all the parameters relevant to plant growth (aerial fresh weight, aerial dry weight, root dry weight, plant height, and leaf area) were significantly affected with rising Cr(VI) concentrations; nitrogen addition partly alleviated Cr(VI) toxic effects. Cr(VI) is known to impede several processes essential for plant growth such as photosynthesis, mineral uptake, enzyme and gene function, that inevitably result in reduced plant growth [6,11,45]. According to Kale et al. [47], plant growth of hydroponically grown *P. oleracea* was severely affected by increasing Cr contents, while plant accumulated significant amounts of Cr compared to other species (up to 190 mg kg^{-1} dry biomass). Various research articles have reported that nutrient addition to the growth medium alleviated to some degree Cr(VI) stress effects, as was the case with *Arabidopsis thaliana* [48,49]. In these investigations, the effect of N was non-significant, but the experimental settings were very different from ours (i.e., seedlings were watered with nutrient or Cr(VI) solutions in soilless culture). Contrary to such reports, our findings indicate that nitrogen amendment can support the growth of plants under Cr(VI) stress and partially compensate for the negative effects of Cr(VI) on plant physiological and metabolic processes. Leaf growth characteristics were proposed as bio-indicators of heavy metal stress. Cr stress is known to result in reduced leaf area, leaf size, and total leaf number per plant [2,45,50,51]. In the present experiment, a series of parameters such as leaf area and the ratio of leaf weight/total aerial weight were significantly affected with increasing Cr(VI) concentrations. Nitrogen amendment partly alleviated the effects of Cr(VI) stress. In parallel to the leaf growth restriction, significantly lower water content was noticed in above ground tissues. These results are in accordance with other works supporting that toxic effects of Cr(VI) in root tissues, alterations on the membrane structure of stomatal guard cells and the reduced diameter of tracheary vessels under Cr(VI) stress are the main factors that limit the water supply to aboveground tissues and therefore inhibit plant growth [2,51]. It seems that root tissues are the most affected plant parts since heavy metals are usually accumulated in higher amounts in roots compared to other plants parts. This was also the case in the study of Kale et al. [47], who recorded higher amounts of Cr in root tissues, followed by a reduction in root length with increasing Cr content in nutrient solution. Similar results were reported by Dwivedi et al. [52] who evaluated two *Portulaca* species (e.g., *P. tuberosa* and *P. oleracea*) for their phytoremediation capacity of multiple heavy metals (e.g., Cu, Ni, Hg, and Pb) and suggested that roots accumulated the highest amounts of metals, followed by stems, leaves and flowers, regardless of the studied metal. Based on the findings of Anandi et al. [53], this selective accumulation of heavy metals in plant tissues could be due to differences in tolerance to toxic effects, as aerial tissues are more susceptible to stress than roots.

According to the literature, *Portulaca* species have been reported for the phytoremediation of heavy metal-polluted soils, since the species seems to be tolerant to toxic effects of increased contents of various metals. For example, Deepa et al. [54] suggested the efficiency of *P. oleracea* stem cuttings in removing Cu from two different types of soils (e.g., Alfisol and Vertisol), while plant uptake was higher for the Alfisol due to the lower availability of Cu in this particular soil type. Moreover, it is worth to highlight the potency of *Portulaca* species to hyperaccumulate different heavy metals, e.g., Cd, As and Cr; this indicates the presence of efficient defense mechanisms that alleviate heavy metal toxic effects [55]. The suggested mechanisms for stress alleviation include the biosynthesis of osmoregulators such as proline or the induction of antioxidant enzymes, e.g., guaiacol peroxidase (GPX) [47]. Finally Yang et al. [56] reported that purslane above and below ground plant parts showed a very high concentration in various trace elements, including chromium, and further suggested the use of the species as a potential biomonitor or phytoremediator.

Cr(VI) also limits the uptake of N, P, K, Mn, Fe, Cu, Zn and S. It is referred that Cr(VI) root uptake is mainly performed by phosphate and sulphate transporters due to the structural similarity of Cr(VI) to phosphate and sulfate ions [45,57]. Results of de Oliveira et al. [58] indicated that increasing Cr(VI) concentrations resulted in higher sulfate root uptake and elevated sulfur in aerial plant tissues. Elevated phosphorus plant tissue content exerts positive effects on enzymes involved in Cr(VI) reduction [59] and increased P uptake under Cr(VI) stress was noticed in *Citrullus vulgaris* [60]. In *Arabidopsis thaliana* seedlings, high phosphorus concentrations in plant tissues resulted in significantly reduced Cr(VI) tissue concentrations [48]. On the other hand, *Brassica napus* plants subjected to oxidative stress recorded higher phosphorus cell content and results indicated that elevated P content resulted in lower ROS stress [61]. These results are in accordance with the present experimental results, where purslane seemed to selectively absorb soil phosphorus when under Cr(VI) stress and nitrogen amendment had a positive effect on P accumulation in purslane aerial tissues with increasing Cr(VI) concentrations.

5. Conclusions

- Cr(VI) contents in root tissues were orders of magnitude higher than the concentrations found in aerial plant tissues.
- All physiological and growth parameters measured were severely affected and nitrogen in all cases resulted even partially in Cr(VI) stress alleviation.
- Under Cr(VI) stress purslane plants selectively accumulated phosphorus in aerial plant tissues.
- Cr(VI) stress resulted in lower water content in aerial plant tissues.
- Added N did not result in increased Cr(VI) content in aerial biomass compared to same Cr(VI)-amended treatments without N; however, the fact that added N improved plant's growth and physiological functions even when exposed to high Cr(VI) soil concentrations, means that sufficient N fertilization may be a satisfactory treatment to increased purslane tolerance against Cr(VI) toxicity.

- On the same lines, added N makes purslane a species to be further considered for phytoremediation of Cr(VI)-laden soils; however, we acknowledge that more research is necessary before conclusive decisions may be drawn.

Author Contributions: Conceptualization, V.A. and S.A.P.; methodology, V.A. and S.A.P.; software, G.T. and E.N.; validation, G.T. and E.N.; formal analysis, V.A. and S.A.P.; investigation, G.T. and E.N.; resources, G.T. and E.N.; data curation, G.T. and E.N.; writing—original draft preparation, G.T. and E.N.; writing—review and editing, V.A. and S.A.P.; visualization, V.A. and S.A.P.; supervision, V.A. and S.A.P.; project administration, V.A.; funding acquisition, V.A. and S.A.P. All authors have read and agreed to the published version of the manuscript.

Funding: We acknowledge the funding of this research by Greece and the European Union (European Social Fund- ESF) through the Operational Programme "Human Resources Development, Education and Lifelong Learning 2014–2020" in the context of the project "Study of cultivation methods for purslane (*Portulaca oleracea* L.) as a plant species for remediation of polluted soils" (MIS 5048931). This funding is highly appreciated.

Institutional Review Board Statement: Not applicable.

Informed Consent Statement: Not applicable.

Data Availability Statement: The study did not report any data.

Acknowledgments: This research is co-financed by Greece and the European Union (European Social Fund-ESF) through the Operational Programme «Human Resources Development, Education and Lifelong Learning 2014–2020» in the context of the project "Study of cultivation methods for purslane (*Portulaca oleracea* L.) as a plant species for remediation of polluted soils" (MIS 5048931).

Conflicts of Interest: The authors declare no conflict of interest.

References

1. Garrett, R.G. Natural sources of metals to the environment. *Hum. Ecol. Risk Assess.* **2000**, *6*, 945–963. [CrossRef]
2. Shanker, A.K.; Cervantes, C.; Loza-Tavera, H.; Avudainayagam, S. Chromium toxicity in plants. *Environ. Int.* **2005**, *31*, 739–753. [CrossRef]
3. Lilli, M.A.; Moraetis, D.; Nikolaidis, N.P.; Karatzas, G.P.; Kalogerakis, N. Characterization and mobility of geogenic chromium in soils and river bed sediments of Asopos basin. *J. Hazard. Mater.* **2015**, *281*, 12–19. [CrossRef]
4. Megremi, I.; Vasilatos, C.; Vassilakis, E.; Economou-Eliopoulos, M. Spatial diversity of Cr distribution in soil and groundwater sites in relation with land use management in a Mediterranean region: The case of C. Evia and Assopos-Thiva Basins, Greece. *Sci. Total Environ.* **2019**, *651*, 656–667. [CrossRef]
5. Jobby, R.; Jha, P.; Yadav, A.K.; Desai, N. Biosorption and biotransformation of hexavalent chromium [Cr(VI)]: A comprehensive review. *Chemosphere* **2018**, *207*, 255–266. [CrossRef] [PubMed]
6. Shahid, M.; Shamshad, S.; Rafiq, M.; Khalid, S.; Bibi, I.; Niazi, N.K.; Dumat, C.; Rashid, M.I. Chromium speciation, bioavailability, uptake, toxicity and detoxification in soil-plant system: A review. *Chemosphere* **2017**, *178*, 513–533. [CrossRef] [PubMed]
7. US EPA. Chromium Compounds: Hazard Summary. Available online: https://www.epa.gov/sites/production/files/2016-09/documents/chromium-compounds.pdf (accessed on 15 June 2021).
8. Kumar, P.; Tokas, J.; Singal, H.R. Amelioration of Chromium VI Toxicity in Sorghum (*Sorghum bicolor* L.) using Glycine Betaine. *Sci. Rep.* **2019**, *9*, 1–15. [CrossRef] [PubMed]
9. Kierczak, J.; Pietranik, A.; Pędziwiatr, A. Ultramafic geoecosystems as a natural source of Ni, Cr, and Co to the environment: A review. *Sci. Total Environ.* **2021**, *755*. [CrossRef] [PubMed]
10. Antoniadis, V.; Levizou, E.; Shaheen, S.M.; Ok, Y.S.; Sebastian, A.; Baum, C.; Prasad, M.N.V.; Wenzel, W.W.; Rinklebe, J. Trace elements in the soil-plant interface: Phytoavailability, translocation, and phytoremediation—A review. *Earth Sci. Rev.* **2017**, *171*, 621–645. [CrossRef]
11. Ertani, A.; Mietto, A.; Borin, M.; Nardi, S. Chromium in Agricultural Soils and Crops: A Review. *Water Air Soil Pollut.* **2017**, *228*. [CrossRef]
12. Ahemad, M. Enhancing phytoremediation of chromium-stressed soils through plant-growth-promoting bacteria. *J. Genet. Eng. Biotechnol.* **2015**, *13*, 51–58. [CrossRef]
13. Sharma, I.; Pati, P.K.; Bhardwaj, R. Effect of 28-homobrassinolide on antioxidant defence system in *Raphanus sativus* L. under chromium toxicity. *Ecotoxicology* **2011**, *20*, 862–874. [CrossRef] [PubMed]
14. Ali, S.; Chaudhary, A.; Rizwan, M.; Anwar, H.T.; Adrees, M.; Farid, M.; Irshad, M.K.; Hayat, T.; Anjum, S.A. Alleviation of chromium toxicity by glycinebetaine is related to elevated antioxidant enzymes and suppressed chromium uptake and oxidative stress in wheat (*Triticum aestivum* L.). *Environ. Sci. Pollut. Res.* **2015**, *22*, 10669–10678. [CrossRef] [PubMed]

15. Antoniadis, V.; Zanni, A.A.; Levizou, E.; Shaheen, S.M.; Dimirkou, A.; Bolan, N.; Rinklebe, J. Modulation of hexavalent chromium toxicity on *Origanum vulgare* in an acidic soil amended with peat, lime, and zeolite. *Chemosphere* **2018**, *195*, 291–300. [CrossRef] [PubMed]

16. Srivastava, S.; Srivastava, S.; Prakash, S.; Srivastava, M.M. Fate of trivalent chromium in presence of organic acids: A hydroponic study on the tomato plant. *Chem. Speciat. Bioavailab.* **1998**, *10*, 147–150. [CrossRef]

17. Ding, M.; Shi, X. Molecular mechanisms of Cr(VI)-induced carcinogenesis. *Mol. Cell. Biochem.* **2002**, *234–235*, 293–300. [CrossRef]

18. Skeffington, R.A.; Shewry, P.R.; Peterson, P.J. Chromium Uptake and Transport in Barley Seedlings (*Hordeum vulgare* L.). *Planta* **1976**, *132*, 209–214. [CrossRef]

19. Prasad, S.; Yadav, K.K.; Kumar, S.; Gupta, N.; Cabral-Pinto, M.M.S.; Rezania, S.; Radwan, N.; Alam, J. Chromium contamination and effect on environmental health and its remediation: A sustainable approaches. *J. Environ. Manag.* **2021**, *285*, 112174. [CrossRef]

20. Nikalje, G.C.; Saini, N.; Suprasanna, P. Halophytes and Heavy Metals: Interesting Partnerships. In *Plant-Metal Interactions*; Srivastava, S., Srivastava, A., Suprasanna, P., Eds.; Springer Nature: Cham, Switzerland, 2019; pp. 99–118. ISBN 9783030207328.

21. Karkanis, A.C.; Petropoulos, S.A. Physiological and growth responses of several genotypes of common purslane (*Portulaca oleracea* L.) under Mediterranean semi-arid conditions. *Not. Bot. Horti Agrobot. Cluj-Napoca* **2017**, *45*, 569–575. [CrossRef]

22. Petropoulos, S.A.; Karkanis, A.; Martins, N.; Ferreira, I.C.F.R. Edible halophytes of the Mediterranean basin: Potential candidates for novel food products. *Trends Food Sci. Technol.* **2018**, *74*, 69–84. [CrossRef]

23. Nikalje, G.C.; Suprasanna, P. Coping with metal toxicity—Cues from halophytes. *Front. Plant Sci.* **2018**, *9*, 1–11. [CrossRef]

24. Petropoulos, S.; Levizou, E.; Ntatsi, G.; Fernandes, Â.; Petrotos, K.; Akoumianakis, K.; Barros, L.; Ferreira, I. Salinity effect on nutritional value, chemical composition and bioactive compounds content of *Cichorium spinosum* L. *Food Chem.* **2017**, *214*, 129–136. [CrossRef]

25. Lutts, S.; Lefèvre, I. How can we take advantage of halophyte properties to cope with heavy metal toxicity in salt-affected areas? *Ann. Bot.* **2015**, *115*, 509–528. [CrossRef]

26. Lefèvre, I.; Marchal, G.; Meerts, P.; Corréal, E.; Lutts, S. Chloride salinity reduces cadmium accumulation by the Mediterranean halophyte species *Atriplex halimus* L. *Environ. Exp. Bot.* **2009**, *65*, 142–152. [CrossRef]

27. Corrêa, R.C.; Di Gioia, F.; Ferreira, I.C.; Petropoulos, S.A. Wild greens used in the Mediterranean diet. In *The Mediterranean Diet: An Evidence-Based Approach*; Preedy, V., Watson, R., Eds.; Academic Press: London, UK, 2020; pp. 209–228. ISBN 9788578110796.

28. Correa, R.C.G.; Di Gioia, F.; Ferreira, I.; SA, P. Halophytes for Future Horticulture: The Case of Small-Scale Farming in the Mediterranean Basin. In *Halophytes for Future Horticulture: From Molecules to Ecosystems towards Biosaline Agriculture*; Grigore, M.-N., Ed.; Springer Nature: Cham, Switzerland, 2020; pp. 1–28. ISBN 9783030178543.

29. Rodriguez, E.; Santos, C.; Azevedo, R.; Moutinho-Pereira, J.; Correia, C.; Dias, M.C. Chromium (VI) induces toxicity at different photosynthetic levels in pea. *Plant Physiol. Biochem.* **2012**, *53*, 94–100. [CrossRef] [PubMed]

30. Nagarajan, M.; Ganesh, K.S. Effect of Chromium on Growth, Biochemicals and Nutrient Accumulation of Paddy (*Oryza sativa* L.). *Int. Lett. Nat. Sci.* **2014**, *23*, 63–71. [CrossRef]

31. Moral, R.; Navarro Pedreno, J.; Gomez, I.; Mataix, J. Effects of chromium on the nutrient element content and morphology of tomato. *J. Plant Nutr.* **1995**, *18*, 815–822. [CrossRef]

32. Kumar, S.; Joshi, U.N. Nitrogen metabolism as affected by hexavalent chromium in sorghum (*Sorghum bicolor* L.). *Environ. Exp. Bot.* **2008**, *64*, 135–144. [CrossRef]

33. Alyazouri, A.; Jewsbury, R.; Tayim, H.; Humphreys, P.; Al-Sayah, M.H. Applicability of Heavy-Metal Phytoextraction in United Arab Emirates: An Investigation of Candidate Species. *Soil Sediment Contam.* **2014**, *23*, 557–570. [CrossRef]

34. Feller, U.; Kopriva, S.; Vassileva, V. Plant nutrient dynamics in stressful environments: Needs interfere with burdens. *Agriculture* **2018**, *8*, 97. [CrossRef]

35. Haneklaus, S.H.; Bloem, E.; Schnug, E. Hungry plants—A short treatise on how to feed crops under stress. *Agriculture* **2018**, *8*, 43. [CrossRef]

36. Subpiramaniyam, S. *Portulaca oleracea* L. for phytoremediation and biomonitoring in metal-contaminated environments. *Chemosphere* **2021**, *280*, 130784. [CrossRef] [PubMed]

37. Sdouga, D.; Ben Amor, F.; Ghribi, S.; Kabtni, S.; Tebini, M.; Branca, F.; Trifi-Farah, N.; Marghali, S. An insight from tolerance to salinity stress in halophyte *Portulaca oleracea* L.: Physio-morphological, biochemical and molecular responses. *Ecotoxicol. Environ. Saf.* **2019**, *172*, 45–52. [CrossRef]

38. Mittler, R. Abiotic stress, the field environment and stress combination. *Trends Plant Sci.* **2006**, *11*, 15–19. [CrossRef]

39. Alyazouri, A.; Jewsbury, R.; Tayim, H.; Humphreys, P.; Al-Sayah, M.H. Uptake of Chromium by *Portulaca oleracea* from Soil: Effects of Organic Content, pH, and Sulphate Concentration. *Appl. Environ. Soil Sci.* **2020**, *2020*. [CrossRef]

40. Rahbarian, R.; Azizi, E.; Behdad, A.; Mirblook, A. Effects of Chromium on Enzymatic/Nonenzymatic Antioxidants and Oxidant Levels of *Portulaca oleracea* L. *J. Med. Plants Prod.* **2019**, *1*, 21–31.

41. Renna, M.; Cocozza, C.; Gonnella, M.; Abdelrahman, H.; Santamaria, P. Elemental characterization of wild edible plants from countryside and urban areas. *Food Chem.* **2015**, *177*, 29–36. [CrossRef] [PubMed]

42. Brozou, E.; Ioannou, Z.; Dimirkou, A. Removal of Cr (VI) and Cr (III) From Polluted Water and Soil Sown with Beet (*Beta vulgaris*) or Celery (*Apium graveolens*) after the Addition of Modified Zeolites. *Int. J. Waste Resour.* **2018**, *8*. [CrossRef]

43. Koutroubas, S.D.; Antoniadis, V.; Damalas, C.A.; Fotiadis, S. Sunflower growth and yield response to sewage sludge application under contrasting water availability conditions. *Ind. Crops Prod.* **2020**, *154*, 112670. [CrossRef]

44. Jones, J.B.J.; Case, V.W. Chapter 15 Sampling, Handling, and Analyzing. In *Soil Testing and Plant Analysis*; Westerman, R.L., Ed.; Soil Science Society of America Inc.: Madison, WI, USA, 1990; Volume 3, pp. 389–427. ISBN 0891187936.
45. Oliveira, H. Chromium as an Environmental Pollutant: Insights on Induced Plant Toxicity. *J. Bot.* **2012**, *2012*, 375843. [CrossRef]
46. Appenroth, K.J.; Keresztes, Á.; Sárvári, É.; Jaglarz, A.; Fischer, W. Multiple effects of chromate on *Spirodela polyrhiza*: Electron microscopy and biochemical investigations. *Plant Biol.* **2003**, *5*, 315–323. [CrossRef]
47. Kale, R.A.; Lokhande, V.H.; Ade, A.B. Investigation of chromium phytoremediation and tolerance capacity of a weed, *Portulaca oleracea* L. in a hydroponic system. *Water Environ. J.* **2015**, *29*, 236–242. [CrossRef]
48. López-Bucio, J.; Hernández-Madrigal, F.; Cervantes, C.; Ortiz-Castro, R.; Carreón-Abud, Y.; Martínez-Trujillo, M. Phosphate relieves chromium toxicity in *Arabidopsis thaliana* plants by interfering with chromate uptake. *BioMetals* **2014**, *27*, 363–370. [CrossRef]
49. Castro, R.O.; Trujillo, M.M.; López Bucio, J.; Cervantes, C.; Dubrovsky, J. Effects of dichromate on growth and root system architecture of *Arabidopsis thaliana* seedlings. *Plant Sci.* **2007**, *172*, 684–691. [CrossRef]
50. Levizou, E.; Zanni, A.A.; Antoniadis, V. Varying concentrations of soil chromium (VI) for the exploration of tolerance thresholds and phytoremediation potential of the oregano (*Origanum vulgare*). *Environ. Sci. Pollut. Res.* **2019**, *26*, 14–23. [CrossRef]
51. Chatterjee, J.; Chatterjee, C. Phytotoxicity of cobalt, chromium and copper in cauliflower. *Environ. Pollut.* **2000**, *109*, 69–74. [CrossRef]
52. Dwivedi, S.; Mishra, A.; Kumar, A.; Tripathi, P.; Dave, R.; Dixit, G.; Tiwari, K.K.; Srivastava, S.; Shukla, M.K.; Tripathi, R.D. Bioremediation potential of genus *Portulaca* L. collected from industrial areas in Vadodara, Gujarat, India. *Clean Technol. Environ. Policy* **2012**, *14*, 223–228. [CrossRef]
53. Anandi, S.; Thangavel, P.; Subburam, V. Influence of aluminium on the restoration potential of a terrestrial vascular plant, *Portulaca oleracea* L. as a biomonitoring tool of fresh water aquatic environments. *Environ. Monit. Assess.* **2002**, *78*, 19–29. [CrossRef] [PubMed]
54. Deepa, R.; Senthilkumar, P.; Sivakumar, S.; Duraisamy, P.; Subbhuraam, C.V. Copper availability and accumulation by *Portulaca oleracea* Linn. stem cutting. *Environ. Monit. Assess.* **2006**, *116*, 185–195. [CrossRef] [PubMed]
55. Tiwari, K.K.; Dwivedi, S.; Mishra, S.; Srivastava, S.; Tripathi, R.D.; Singh, N.K.; Chakraborty, S. Phytoremediation efficiency of *Portulaca tuberosa* rox and *Portulaca oleracea* L. naturally growing in an industrial effluent irrigated area in Vadodra, Gujrat, India. *Environ. Monit. Assess.* **2008**, *147*, 15–22. [CrossRef]
56. Yang, Y.; Zhou, Z.; Bai, Y.; Jiao, W.; Chen, W. Trace Elements in Dominant Species of the Fenghe River, China: Their Relations to Environmental Factors. *J. Environ. Qual.* **2016**, *45*, 1252–1258. [CrossRef]
57. De Oliveira, L.M.; Ma, L.Q.; Santos, J.A.G.; Guilherme, L.R.G.; Lessl, J.T. Effects of arsenate, chromate, and sulfate on arsenic and chromium uptake and translocation by arsenic hyperaccumulator *Pteris vittata* L. *Environ. Pollut.* **2014**, *184*, 187–192. [CrossRef] [PubMed]
58. De Oliveira, L.M.; Gress, J.; De, J.; Rathinasabapathi, B.; Marchi, G.; Chen, Y.; Ma, L.Q. Sulfate and chromate increased each other's uptake and translocation in As-hyperaccumulator *Pteris vittata*. *Chemosphere* **2016**, *147*, 36–43. [CrossRef]
59. Ma, L.; Chen, N.; Feng, C.; Li, M.; Gao, Y.; Hu, Y. Coupling enhancement of Chromium(VI) bioreduction in groundwater by phosphorus minerals. *Chemosphere* **2020**, *240*, 124896. [CrossRef]
60. Dube, B.K.; Tewari, K.; Chatterjee, J.; Chatterjee, C. Excess chromium alters uptake and translocation of certain nutrients in citrullus. *Chemosphere* **2003**, *53*, 1147–1153. [CrossRef]
61. Yao, Y.; Sun, H.; Xu, F.; Zhang, X.; Liu, S. Comparative proteome analysis of metabolic changes by low phosphorus stress in two *Brassica napus* genotypes. *Planta* **2011**, *233*, 523–537. [CrossRef] [PubMed]

 *horticulturae*

Article

Altered Carbohydrate Allocation Due to Soil Water Deficit Affects Summertime Flowering in Meiwa Kumquat Trees

Naoto Iwasaki *, Asaki Tamura and Kyoka Hori

School of Agriculture, Meiji University, Kawasaki, Kanagawa 214-8571, Japan; a.tamura1995@gmail.com (A.T.); kyohori66@gmail.com (K.H.)
* Correspondence: iwasaki@meiji.ac.jp; Tel.: +81-44-934-7817

Received: 18 July 2020; Accepted: 24 August 2020; Published: 26 August 2020

Abstract: The summertime flowers of the ever-flowering Meiwa kumquat (*Fortunella crassifolia* Swingle) are the most useful for fruit production in Japan; however, summertime flowers bloom in three or four successive waves at approximately 10 day intervals, resulting in fruit of different maturity occurring on the same tree. Soil water deficit (SWD) treatment has been shown to reduce the flowering frequency and improve harvest efficiency; therefore, in this study, the effects of SWD treatment on the accumulation of soluble sugars in each tree organ above-ground were examined and it was discussed how SWD affects the whole-tree water relations and sugar accumulation by osmoregulation. The number of first-flush summertime flowers was higher in SWD-treated trees than non-treated control (CONT) trees (177.0 and 58.0 flowers, respectively), whereas the second- and third-flush flowers were only observed in CONT trees. The soluble sugar content was higher in SWD treated trees than CONT trees for all organs and tended to be higher in current-year organs than previous-year organs; however, when the sugar content of the current-year spring stems exceeded approximately 100 mg g^{-1} dry weight, the current-year leaf water potential decreased sharply and the rate of increase in the number of first-flush flowers also tended to decrease. SWD treatment significantly increased the total sugar content of the xylem tissue of the scaffold branches to three times the value in CONT trees ($p = 0.001$); however, the increase was observed even in sucrose, a disaccharide, similar to that in monosaccharides such as glucose and fructose. These results suggest that the increased sugar levels in the xylem tissue resulted from not only osmoregulation but also other factors as well; therefore, these sugars may affect whole-tree water relations as well as the development of flower buds.

Keywords: osmoregulation; sugar accumulation; water stress; xylem tissue

1. Introduction

Meiwa kumquat (*Fortunella crassifolia* Swingle) is an ever-flowering tree that usually blooms in spring (April), summer (June to August), and autumn (September) in Japan. The spring flowers appear on the previous year shoots, but are very few in number and not usually used for fruit production. By contrast, the summertime flowers appear on each axil of the current spring shoots that emerge around May and are the most useful for fruit production [1]; therefore, the flower bud differentiation period of the summertime flowers is regarded as approximately 1 month from the termination of spring shoot elongation to flowering [2]; however, the flowering of summertime flowers is divided into three to four separate time points, and when flowers do not bear fruitlets, new flowers appear at the same axil after 10–14 days. The resulting flowers are known as the first-, second-, and third-flush flowers in Japanese production areas [1,3]. The fruits from first-flush flowers tend to grow well and mature early. The fruit quality also tends to be good in first-flush flowers. Since there are few first-flush flowers—the number of first-flush flowers also varies annually—the second-flush flowers are used for fruit production in general [1].

Many previous studies on Meiwa kumquat have shown that the application of a soil water deficit (SWD) for 2–4 weeks during the flower bud differentiation period increases the number of first-flush flowers [4–11]; however, the optimal duration of SWD treatment and its effectiveness varies between years. Iwasaki et al. [7] found that severe drought stress that reduced the predawn leaf water potential to below −1.7 MPa or increased the leaf abscisic acid (ABA) content to above 5 nmol g^{-1} dry weight (DW) did not increase the number of first-flush flowers, suggesting that the leaf water potential needs to be maintained between 0 and −1.7 MPa; however, this condition is difficult to achieve because the leaf water potential fluctuates not only with soil water content but also with the relative humidity of the atmosphere [6,11].

Iwasaki and Hiratsuka [8] reported that the leaf water potential tends to be lower in Meiwa kumquat trees with larger trunk and scaffold branches under drought conditions. They also noted that transpiration from the trunk surface does not change despite rapid decreases in leaf transpiration under drought conditions, resulting in continuous water loss from the trunk at approximately one-sixth of the leaf transpiration rate. In some woody species, water stored in the trunk is transported to organs such as the leaves when required, but this process does not occur in isohydric plants [12–14]. Isohydric plants maintain a constant midday leaf water potential under both abundant water and drought conditions by reducing stomatal conductance to limit transpiration when required [14,15]. Thus, the kumquat tree is considered isohydric because SDW treatment rapidly decreases the transpiration rate in the leaves [9], and the water stored in the trunk does not migrate to the leaves. Plants that have encountered drought actively accumulate soluble sugars and free amino acids to prevent cell dehydration, via a process known as osmoregulation [16]. Thus the migration of water between the organs within a tree may be related to the amount of soluble sugars that have accumulated in that organ.

In the present study, the accumulation of soluble sugars in the organs of Meiwa kumquat trees under SWD treatment was investigated and considered the relationship between the sugar content of each organ and the number of first-flush flowers.

2. Materials and Methods

2.1. Plant Materials and Treatment

Six eighteen-year-old Meiwa kumquat trees that had been grafted onto trifoliate orange [*Poncirus trifiliata* (L.) Raf.] were used in the experiment, which was carried out during the growing season in 2016. The trees were planted in 29-L non-woven fabric pots containing commercially available humus-rich soil and each tree had a height and canopy diameter of approximately 1.2 and 1.5 m, respectively. Bud burst and elongation of spring shoots began around 25 April 2016, and were completed around 20 May 2016.

Three trees were allocated to each of the SWD treatment and non-treated control (CONT) groups on June 4 and maintained for 12 days in a greenhouse at Meiji University (Kawasaki, Kanagawa, Japan; 35°61′ N, 139°55′ E). For the SWD treatment, the soil water content was reduced by withholding irrigation, and a small volume of water (approximately 300 mL) was applied when the soil water content fell to <15%. During the treatment, the soil water content was monitored at a depth of approximately 10 cm between the trunk and the pot rim using an ECH$_2$O soil moisture sensor (EC-5) with an Em50 digital data logger (METER Group, Inc., Pullman, WA, USA). The leaf water potential was determined on June 8, 11, and 14 using a psychrometer (WP4-T; METER Group, Inc., Pullman, WA, USA) to monitor the severity of water stress in the trees. One matured leaf in the middle of two current shoots per tree was collected at 9:00 a.m. on each measurement day.

2.2. Observation of Flowering Behavior and Determination of Total Soluble Sugars and Starch

Ten current-year spring shoots of approximately 10 cm length were selected from each tree to observe flowering behavior. The number of flowers that opened on each shoot was recorded each day from 20 June to 31 July, and the number of first-flush flowers and the total number of flowers per 10 shoots per tree was counted. One scaffold branch per tree was harvested on 15 July at the end of SWD treatment, and the plant parts were categorized into current-year spring leaves (current leaves), current-year spring stems (current stems), previous-year leaves (previous leaves), previous-year stems (previous stems), and scaffold branch, which included branches >2 years old. The scaffold branch was then further divided into bark and wood (xylem tissue). The fresh weight of each organ was measured, following which the organs were stored at −70 °C for further analysis. To analyze the soluble sugar and starch contents of the organs, the samples were freeze-dried and ground to powder. Soluble sugars were then extracted by adding 0.1 g of dried sample to 50 mL of 80% ethanol and leaving for 24 h at room temperature, following which the ethanol was removed from the extract using a rotary evaporator (N-1000; Tokyo Rika Kikai Co., Tokyo, Japan) at 35 °C and the remaining aqueous phase was diluted to 50 mL. The total sugar content in 1 mL of the diluted sample was then determined using the anthrone method. The starch content of the organs was determined with the starch-iodine colorimetric method [17] using the alcohol-insoluble solids (AIS) that were obtained during sugar extraction. The soluble sugar and starch contents of each plant part were expressed as mg g^{-1} dry weight (DW). The water content (WC) of each plant part was calculated as WC = (FW − DW)/FW where FW is the fresh weight.

2.3. Sugar Composition of the Xylem Tissue and Bark Extracts of the Scaffold Branch

Following the extraction of sugars from each dried sample as described above, the aqueous phase that remained after ethanol removal was lyophilized. The residue was then dissolved in 2 mL of ultrapure water to prepare an analytical sample, which was filtered using a 0.45 μm membrane filter before being injected into a high-performance liquid chromatography (HPLC) system (LC-2000Plus series, JASCO, Tokyo, Japan). Sugars were separated in a Shodex SC1011 column (Shodex SC1011; Showa Denko K.K., Tokyo) that was maintained at 80 °C using ultrapure water as the mobile phase with a flow rate of 1.0 mL min^{-1}. Sucrose, glucose, and fructose were then identified and quantified by comparison with peaks of known standards using a refractive index detector (RI-2031; JASCO, Tokyo, Japan).

2.4. Statistical Analyses

Each treatment was performed on three trees (one tree per plot) and differences between the treatments were assessed using *t*-tests. The differences in sugar and starch content among the organs were compared using the Tukey HSD test; the coefficients of correlation and determination were obtained by regression analysis. Percentage values were arcsine transformed before analysis. Differences between the means were analyzed using the software KaleidaGraph v.4.5 (Synergy Software, Reading, PA, USA) when ANOVA was significant at $p = 0.05$ or $p = 0.01$, whereas BellCurve for Excel, v.2.00 (Social Survey Research Information Co., Ltd. Tokyo, Japan) was used for regression analysis.

3. Results

3.1. Changes in the Soil Water Content and Leaf Water Potential during SWD Treatment

The mean soil water content gradually decreased from 32.7% on June 8 to 12.1% on June 14 in the SWD treatment group but remained relatively constant in the CONT group (Table 1). The average leaf water potential decreased from −1.9 MPa on June 8 to −2.7 MPa on June 14 in the SWD-treated trees but remained at −1.6 to −1.7 MPa in the CONT trees.

Table 1. Changes in the soil water content and the leaf water potential at mid-morning during soil water deficit (SWD) treatment.

	8 Jun	11 Jun	14 Jun
Soil water content (%)			
CONT	41.6 [z]	39.2	45.1
SWD	32.7	9.3	12.1
Significance	ns [y]	*	*
Leaf water potential (Mpa)			
CONT	−1.75	−1.61	−1.62
SWD	−1.92	−2.38	−2.73
Significance	ns	*	*

[z] Measurements were undertaken at 9:00 AM and values are mean of 3 replications. [y] ns and * indicate a non-significant and significant differences between the treatments at $p = 0.05$ by *t*-test, respectively.

3.2. Effect of SWD Treatment on Flowering Behavior

Blooming of the first- to fourth-flush flowers was observed by the end of July on the CONT trees, whereas only first-flush flowers were observed on the SWD-treated trees (Figure 1). The SWD-treated trees had a higher maximum number of flowers open per day than the CONT trees (46 and 26 flowers, respectively) and a significantly higher number of first-flush flowers (177.0 and 58.0 flowers per 10 shoots, respectively) (Table 2). The SWD-treated trees also had a higher total number of flowers than the CONT trees, though this difference was not significant.

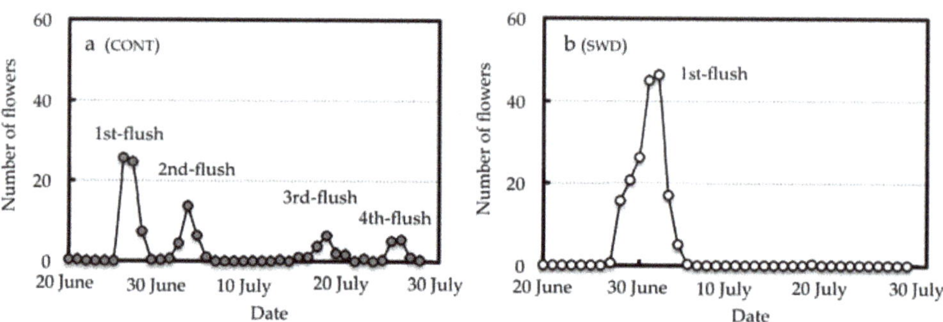

Figure 1. Difference of flowering behavior between control (CONT) (**a**) and soil water deficit (SWD) (**b**) in Meiwa kumquat trees. The number of flowers means the number of flowers that bloomed per day in 10 shoots per tree. Each plot indicates the mean of three replications.

Table 2. Effects of SWD treatment on the number of first-flush flowers and total flowers.

	1st-Flush Flowers [z]	Total Flowers
CONT	58.0 [y]	113.7
SWD	177.0	177.3
Significance	* [x]	ns

[z] Flowers bloomed from June 26 to June 29 in CONT, June 27 to July 5 in SWD. [y] Number of flowers appeared 10 shoots per tree ($n = 3$). [x] * indicate significant difference between treatments at $p = 0.05$ by *t*-test.

3.3. Effects of SWD Treatment on the Soluble Sugar, Starch and Water Contents of the Plant Parts

For all plant parts, the soluble sugar content was significantly higher in the SWD-treated trees than in the CONT trees (Figure 2). In CONT trees, the soluble sugar content was significantly higher in the current leaves and stems than in the previous leaves and stems respectively. The xylem tissue of the scaffold branches showed the lowest soluble sugar content among all of the plant parts in the CONT trees but also showed the greatest increase following SWD treatment (3.07 fold increase). By contrast, the bark of the scaffold branches had a higher soluble sugar content but experienced the lowest rate of increase following SWD treatment (1.16 fold increase). The starch content of the organs tended to be lower in the SWD treated trees, compared with the CONT trees, except for the current and previous leaves; however, this difference was only significant for the current stems. In the CONT trees, the starch contents were higher in the previous stems and leaves than in the current stems and leaves, respectively. There was no significant difference in water content between the SWD-treated trees and the CONT trees in any of the plant parts except for the current stems and the bark of the scaffold branches. In the CONT trees, the water content was highest in the current leaves (approximately 73%) and lowest in the xylem tissue of the scaffold branches (approximately 36%).

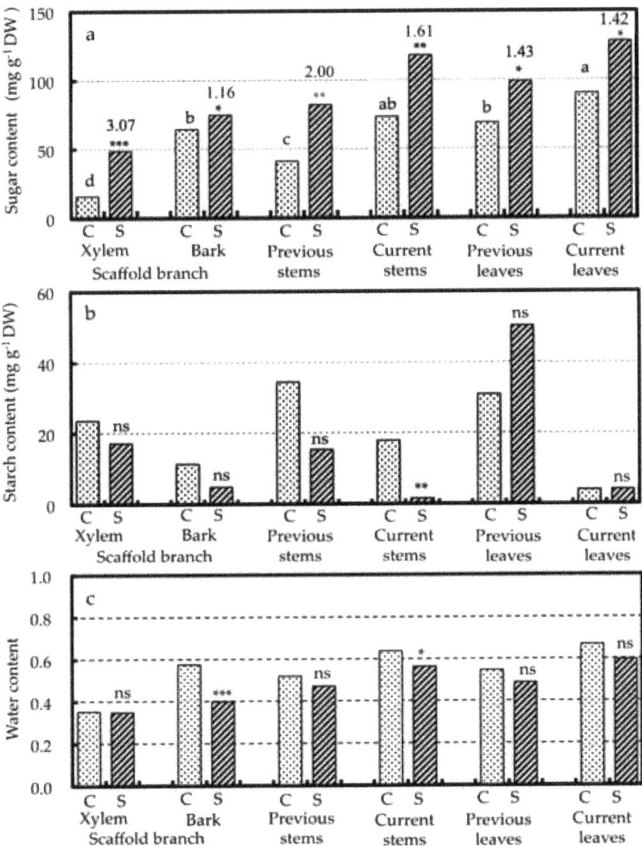

Figure 2. Soluble sugar (**a**), starch (**b**) and water (**c**) contents of various plant parts in Meiwa kumquat trees as affected by SWD treatment. C and S indicate CONT and SWD, respectively. The numbers above the bar of sugar content are the ratio of SWD to CONT. *, **, *** and ns indicate significant difference between the treatments at p = 0.05, 0.01 and 0.001 and non-significant by *t*-test, respectively (n = 3). Different letters above CONT bar indicate significant differences among the plant parts at p = 0.05 by Tukey HSD test (n = 3). Water content is expressed as the ratio of water to fresh weight.

Sucrose, glucose, and fructose were detected in both the xylem tissue and the bark of the scaffold branches (Table 3); however, sucrose was the predominant sugar in both tissues, whereas glucose was only detected at low levels in the bark. The concentrations of all three sugars were significantly higher in the xylem tissue of the SWD-treated trees than in that of the CONT trees, with the SWD treated trees exhibiting a four fold increase in sucrose and an approximately six fold increase in the monosaccharides. A similar pattern was also observed in the bark, although none of these differences were significant.

Table 3. Effect of SWD treatment on sugar compositions of xylem and bark in scaffold branch of Meiwa kumquat.

	Sucrose	Glucose	Fructose	Total
Xylem tissue				
CONT	3.4 [z]	1.4	1.7	6.5
SWD	13.5 (4.0) [y]	9.6 (6.6)	9.8 (5.7)	32.8 (5.0)
significance	* [x]	**	**	**
Bark				
CONT	21.0	0.5	1.7	20.4
SWD	25.1 (1.2)	1.1 (2.1)	2.0 (1.2)	28.3 (1.4)
significance	ns	ns	ns	ns

[z] Values are the mean of 3 replications and expressed by mg g^{-1} DW. [y] Number in parentheses indicate the ratio of SWD to CONT. [x] * and ** indicate significant difference between treatments at $p = 0.05$ and $p = 0.01$, respectively, by t-test.

3.4. Relationships between Sugar Content of the Current Spring Stems and the Number of First-Flush Flowers or Leaf Water Potential of the Current Spring Leaves

The number of first-flush flowers was positively correlated with the sugar content in each tree organ, but significantly negatively correlated with the water potential of the current leaves (Table 4). For both relationships, the correlation coefficients were highest in the current stems ($p = 0.01$), however, when the sugar content of the current stems exceeded approximately 100 mg g^{-1} DW, the current leaf water potential rapidly decreased while the number of first-flush flowers plateaued (Figure 3).

Table 4. Correlation coefficients between sugar contents in each tree organ and the number of first-flush flowers or the current leaf water potential.

	Number of 1st-Flush Flowers	Current Leaf Water Potential
Scaffold branch		
Xylem	0.8385 * [z]	−0.8727 * [z]
Bark	0.8395 *	−0.9011 *
Stems		
Previous year stems	0.8662 *	−0.9565 **
Current-year spring stems	0.9584 **	−0.9567 **
Leaves		
Previous year leaves	0.7015	−0.9400 **
Current-year spring leaves	0.7055	−0.7649

[z] ** and * indicate significant at $p = 0.01$ and $p = 0.05$, respectively.

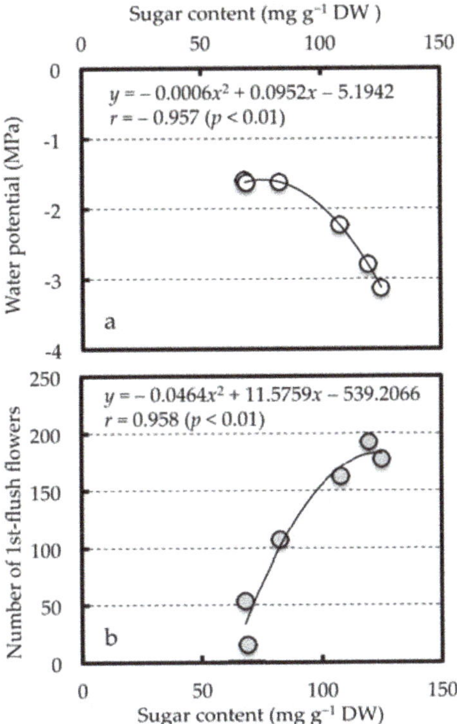

Figure 3. Relations between sugar content of the current shoot and the water potential of current leaves (**a**) or the number of first-flush flowers (**b**). The number of first-flush flowers indicates the number of flowers that opened in 10 current spring shoots per tree.

4. Discussion

In this study, it was found that SWD treatment reduced the frequency of sequential blooming of summertime flowers and increased the number of first-flush summertime flowers in Meiwa kumquat; however, the total number of flowers did not significantly change, indicating that SWD promoted flower bud development rather than differentiation. These results support the findings of previous studies [4,6] and suggest that the application of SWD treatment approximately two weeks after the cessation of spring shoot elongation during flower bud differentiation would accomplish uniform fruit maturation, removing the need to evaluate fruit maturity at harvest and thus simplifying harvesting practices.

In many crops, deficit irrigation is known to increase fruit sugar content without decreasing the yield [18–21], and nowadays, the water stress is often applied to produce high-quality fruits. We found that kumquat trees accumulated considerable amounts of sugars in their scaffold branches, particularly in the xylem tissue, under SWD conditions. Although the scaffold branches had a lower total soluble sugar content per gram dry weight than the leaves, they exhibited a greater proportional increase following SWD treatment compared with the CONT trees (3.1 fold increase in xylem tissue while a 1.4 fold increase in the leaves). In this study, the trunk was divided into bark and wood. The wood is the xylem tissue consists of vessels, tracheid, parenchymal cell, etc., but since the vessels and tracheid are dead cells, sugars cannot be actively absorbed into the cells [22,23]; therefore, we concluded that most sugars are accumulated in the xylem parenchyma cells. In the xylem tissue, sucrose was detected alongside glucose and fructose, and the sucrose content also increased alongside glucose or fructose under SWD conditions. Since the water content of the xylem tissue did not change as a result of SWD treatment, the increase in sugar content might involve not only osmoregulation [16,24] but also other factors. Trees growing in cold areas such as birch are known to accumulate sucrose, raffinose, and stachyose in xylem parenchyma cells during the cold season [25]. These sugars are considered to not only increase the freeze-tolerance of cells but have a stabilizing effect of cell membrane under stress conditions such as freezing, desiccation and high temperature, and that the effect is higher for trisaccharides or disaccharides having a higher molecular weight than monosaccharides [26].

It has been previously shown that the carbohydrate contents of the leaves or bark of shoots that produce the next flowers gradually increase during flower bud differentiation/development period in Satsuma mandarin (*Citrus unshiu*) trees growing outdoors in Valencia, Spain [27], or in early-heating plastic houses in Fukuoka, Japan [28], although they sometimes decrease temporarily; however, none of these studies found a direct relationship between flower bud differentiation and the sugar or starch content. In the present study, the number of first-flush flowers was most closely related to the sugar content of the current shoots that set flowers for fruit production. Similarly, Niii and Okamoto [29] reported that the amount of stored carbohydrates influenced the development of flower buds in Satsuma mandarin, with flower buds showing better development when old leaves existed during budburst. It has also been shown that buds of peach (*Prunus persica*) actively absorbed carbohydrates during dormancy release, which are used for growth metabolism and thereby induce bud development [30]. In addition, Nakajima et al. [31] reported that water stress from early September through December increased the number of 'Tosa Buntan' pomelo (*Citrus grandis*) flowers and caused them to bloom earlier the following spring. Consequently, the authors considered that water-stressed trees were forced into dormancy to promote flower induction, and released from dormancy by water stress in a similar way to the flower buds of coffee (*Coffea arabica*) [32]. SWD treatment may have caused flower bud dormancy to occur earlier in kumquat, although the dormant period, occurring immediately after the cessation of spring shoot elongation, may be short. In recent years, sugars such as sucrose have been reported to be important signals that regulate bud outgrowth and act before hormones in releasing apical dominance [33,34], i.e., sucrose promotes auxin export from the bud, then the axillary buds have sustained outgrowth. Since the flower buds of kumquat are formed in the axils of shoots, the mechanism involved in their development may be similar to that of apical dominance; therefore, the sucrose accumulated in xylem, just as in glucose and fructose, may have promoted the development or budburst of floral buds that originally should have been the second- or third-flush flowers, and made them the first-flush flowers.

Consequently, such sugar accumulation may decrease the water potential of the xylem tissue, resulting in the active accumulation of water due to osmoregulation. Although the sugar content of the trunk was not investigated in this experiment, the sugar content of the scaffold branches is expected to reflect this. Both the trunk and scaffold branches can be regarded as water reservoirs in many tree species [12,35,36]. Nevertheless, the rapid decrease in leaf water potential that was observed during SWD treatment in this experiment indicated that the water supply from the trunk and scaffold branches to the leaves was restricted. Iwasaki et al. [9] reported that the water potential of the trunk xylem tissue was always lower than the leaf water potential in kumquat trees under SWD conditions, which may result from the large accumulation of sugars. It was also found that the relationship between the number of first-flush flowers or current leaf water potential and the sugar content of the current stems changed when the sugar content reached approximately 100 mg g^{-1} DW. Similarly Iwasaki et al. [7] reported that severe water stress expressed by a leaf water potential below -1.7 MPa or a leaf ABA content above 5 nmol g^{-1} did not increase the number of first-flush flowers. Thus, factors related to the level of water stress, such as leaf water potential and sugar content of the plant tissues, appear to have a limited range of influence on the number of first-flush flowers. Although it is necessary to investigate the detailed mechanism of sugar accumulation under drought conditions, the xylem tissue of the trunk and scaffold branches seems to play an important role in the whole-tree water relations through active sugar accumulation.

Author Contributions: Conceived and designed the experiments: N.I.; performed the experiments: A.T. and K.H.; analyzed the data: A.T. and N.I.; wrote the manuscript: A.T. and N.I. All authors have read and agreed to the published version of the manuscript.

Funding: This research received no external funding.

Conflicts of Interest: The authors declare no conflict of interest.

References

1. Hadano, Y. Saibaigijutsu no kiso. Kinkan. In *Kajuenngei Dai Hyakka*; Jyoryoku Tokusan Kaju, Nousangyosonbunka Kyokai: Tokyo, Japan, 2000; Volume 15, pp. 117–119.
2. Abbott, C.E. Blossom-bud differentiation in citrus trees. *Am. J. Bot.* **1935**, *22*, 476–485. [CrossRef]
3. Iwahori, S.; Tominaga, S. Increase in first-flush flowering of 'Meiwa' kumquat, *Fortunella crassifolia* Swingle, trees by paclo-butrazol. *Sci. Hortic.* **1986**, *28*, 347–353. [CrossRef]
4. Iwasaki, N.; Hayasaki, K.; Tanaka, S. Effect of Water Stress on Flowering of Meiwa Kumquat Trees. *Environ. Control Biol.* **2000**, *38*, 105–109. [CrossRef]
5. Iwasaki, N.; Yamaguchi, T. Flowering and Yield of Meiwa Kumquat Trees Are Affected by Duration of Water stress. *Environ. Control Biol.* **2004**, *42*, 241–245. [CrossRef]
6. Iwasaki, N.; Hagiwara, H.; Yasuda, N.; Ono, T. Effects of deficit soil water and deficit vapor-pressure on flowering of Meiwa kumquat trees. *J. SASJ* **2012**, *43*, 8–14.
7. Iwasaki, N.; Nakano, Y.; Suzuki, K.; Mochizuki, A. Relationships between the number of first-flush flowers and leaf water potential or leaf ABA content affected by varying degrees of water stress in Meiwa kumquat (*Fortunella crassifolia* Swingle). *Environ. Control Biol.* **2017**, *55*, 59–64. [CrossRef]
8. Iwasaki, N.; Hiratsuka, S. Effect of trunk size on the relationship between drought stress and first-flush flower number in Meiwa kumquat. *Environ. Control Biol.* **2019**, *57*, 1–7. [CrossRef]
9. Iwasaki, N.; Hori, K.; Ikuta, Y. Xylem plays an important role in regulating the leaf water potential and fruit quality of Meiwa kumquat (*Fortunella crassifolia* Swingle) trees under drought conditions. *Agric. Water Manag.* **2019**, *214*, 47–54. [CrossRef]
10. Ono, T.; Hagiwara, H.; Iwasaki, N. Effects of water stress on leaf water potential, flowering, carbohydrate content and ABA content in Meiwa kumquat trees. *Hort. Res.* **2010**, *9*, 209–213. [CrossRef]
11. Ono, T.; Hagiwara, H.; Yasuda, N.; Takekawa, H.; Iwasaki, N. Relationship between Atmospheric Relative Humidity and Effectiveness of Water Stress on Flowering in Meiwa Kumquat Trees. *Hort. Res.* **2012**, *11*, 81–85. [CrossRef]
12. Chapotin, S.M.; Razanameharizaka, J.H.; Holbrook, N.M. Baobab trees (*Adansonia*) in Madagascar use stored water to flush new leaves but not to support stomatal opening before the rainy season. *New Phytol.* **2006**, *169*, 549–559. [CrossRef] [PubMed]
13. Van den Bilcke, N.; Simbo, D.J.; Samson, R. Water relations and drought tolerance of young African tamarind (*Tamarindus indica* L.) trees. *S. Afr. J. Bot.* **2013**, *88*, 352–360. [CrossRef]
14. Van den Bilcke, N.; Smedt, S.D.; Simbo, D.J.; Samson, R. Sap flow and water use in African baobab (*Adansonia digitata* L.) seedling in response to drought stress. *S. Afr. J. Bot.* **2013**, *88*, 438–446. [CrossRef]
15. Sade, N.; Gebremedhin, A.; Monshelion, M. Risk-taking plants Anisohydric behavior as a stress-resistance trait. *Plant Signal. Behav.* **2012**, *7*, 767–770. [CrossRef] [PubMed]
16. Meyer, R.F.; Boyer, J.S. Osmoregulation, solute distribution, and growth in soybean seedlings having low water potentials. *Planta* **1981**, *151*, 482–489. [CrossRef] [PubMed]
17. Sugiyama, Y.; Ooshiro, A. A simple and rapid analysis of starch content in root and shoot of Satsuma mandarin by using colorimetric determination. *Jpn. J. Soil Sci. Plant Nutr.* **2001**, *72*, 81–84.
18. García-Tejero, I.; Jiménez-Bocanegra, J.A.; Martínez, G.; Romero, R.; Durán-Zuazo, V.H.; Muriel-Fernández, J.L. Positive impact of regulated deficit irrigationon yield and fruit quality in a commercial citrus orchard *Citrus sinensis* (L.) Osbeck cv. Salustiano. *Agric. Water Manag.* **2010**, *97*, 614–622. [CrossRef]
19. González-Altozano, P.; Castel, J.R. Effects of regulated deficit irrigation on "Clementina de Nules" citrus trees. I. Yield and fruit quality effect. *J. Hortic. Sci. Biotech.* **1999**, *74*, 706–713. [CrossRef]
20. Panigrahi, P.; Srivastava, A.K. Effective management of irrigation water in citrus orchards under a water scarce hot sub-humid region. *Sci. Hortic.* **2016**, *210*, 6–13. [CrossRef]

21. Pérez-Pérez, J.; Romero, P.; Navarro, J.; Botía, P. Response of sweet orange cv. 'lane late' to deficit irrigation strategy in two rootstocks. II: Flowering, fruit growth, yield and fruit quality. *Irrig. Sci.* **2008**, *26*, 519–529. [CrossRef]

22. Déjadin, A.; Laurans, F.; Arnaud, D.; Breton, C.; Pilate, G.; Leple, J.C. Wood formation in Angiosperms. *C. R. Biol.* **2010**, *333*, 325–334. [CrossRef] [PubMed]

23. Plomion, C.; Leprovost, G.; Stokes, A. Wood formation in trees. *Plant Physiol.* **2001**, *127*, 1513–1523. [CrossRef] [PubMed]

24. Yakushiji, H.; Nonami, H.; Fukuyama, T.; Ono, S.; Takagi, N.; Hashimoto, Y. Sugar accumulation enhanced by osmoregulation in *Satsuma mandarin* fruit. *J. Am. Soc. Hortic. Sci.* **1996**, *121*, 466–472. [CrossRef]

25. Kasuga, J.; Fujiwara, S.; Arakawa, K. Seasonal changes in accumulation of soluble sugars in birch xylem tissues. *Cryobiol. Cryotechnol.* **2003**, *49*, 185–189.

26. Santarius, K.A. The protective effect of sugars on chloroplast membranes during temperature and water stress and its relationship to frost, desiccation and heat resistance. *Planta* **1973**, *113*, 105–114. [CrossRef]

27. Garcia-Luis, A.; Fornes, F.; Guardiola, J.L. Leaf carbohydrates and flower formation in Citrus. *J. Am. Soc. Hortic. Sci.* **1995**, *120*, 222–227. [CrossRef]

28. Yahata, D.; Oba, Y.; Kuwahara, M. Changes in carbohydrate level, α-amylase activity, indoleacetic acid and gibberellin-like substances in the summer shoots of Wase Satsuma mandarin trees grown indoors during flower-bud differentiation. *J. Jpn. Soc. Hortic. Sci.* **1995**, *64*, 527–533. [CrossRef]

29. Nii, N.; Okamoto, S. Effects of leaf age and defoliation on tree growth, flowering and fruit set of Satsuma mandarin. *J. Jpn. Soc. Hortic. Sci.* **1973**, *42*, 7–12. [CrossRef]

30. Marquat, C.; Vandamme, M.; Gendraud, M.; Petel, G. Dormancy in vegetative buds of peach: Relation between carbohydrate absorption potentials and carbohydrate concentration in the bud during dormancy and its release. *Sci. Hortic.* **1999**, *79*, 151–162. [CrossRef]

31. Nakajima, Y.; Susanto, S.; Hasegawa, K. Influence of water stress in autumn on flower induction and fruiting in young pomelo trees (*Citrus grandis* (L.) Osbeck). *J. Jpn. Soc. Hortic. Sci.* **1993**, *62*, 15–20. [CrossRef]

32. Astegiano, E.D.; Maestri, M.; Estevao, M.D.M. Water stress and dormancy release in flower buds of *Coffea arabica* L.: Water movement into the buds. *J. Hortic. Sci.* **1988**, *63*, 529–533. [CrossRef]

33. Barbier, F.F.; Lunn, J.E.; Beveridge, C.A. Ready, steady, go! A sugar hit starts the race to shoot branching. *Curr. Opin. Plant Biol.* **2015**, *25*, 39–45. [CrossRef] [PubMed]

34. Barbier, F.F.; Dun, E.A.; Kerr, S.C.; Chabikwa, T.G.; Beveridge, C.A. An update on the signals controlling shoot branching. *Trends Plant Sci.* **2019**, *24*, 220–236. [CrossRef] [PubMed]

35. Goldstein, G.; Andrade, J.L.; Meinzer, F.C.; Holbrook, N.M.; Cavelier, J.; Jackson, P.; Celis, A. Stem water storage and diurnal patterns of water use in tropical forest canopy trees. *Plant Cell Environ.* **1998**, *21*, 397–406. [CrossRef]

36. Stratton, L.; Goldstein, G.; Meinzer, F.C. Stem water storage capacity and efficiency of water transport: Their functional significance in a Hawaiian dry forest. *Plant Cell Environ.* **2000**, *23*, 99–106. [CrossRef]

 *horticulturae*

Article

Growth Response of Ginger (*Zingiber officinale*), Its Physiological Properties and Soil Enzyme Activities after Biochar Application under Greenhouse Conditions

Dilfuza Jabborova [1,2,*], Stephan Wirth [2], Mosab Halwani [2], Mohamed F. M. Ibrahim [3], Islam H. El Azab [4], Mohamed M. El-Mogy [5] and Amr Elkelish [6]

[1] Institute of Genetics and Plant Experimental Biology, Uzbekistan Academy of Sciences, Kibray 111208, Uzbekistan

[2] Leibniz Centre for Agricultural Landscape Research (ZALF), Eberswalder Str. 84, 15374 Müncheberg, Germany; swirth@zalf.de (S.W.); mosab.halwanil@zalf.de (M.H.)

[3] Department of Agricultural Botany, Faculty of Agriculture, Ain Shams University, Cairo 11566, Egypt; Ibrahim_mfm@agr.asu.edu.eg

[4] Food Science & Nutrition Department, College of Science, Taif University, P.O. box 11099, Taif 21944, Saudi Arabia; i.helmy@tu.edu.sa

[5] Vegetable Crops Department, Faculty of Agriculture, Cairo University, Giza 12613, Egypt; elmogy@agr.cu.edu.eg

[6] Botany Department, Faculty of Science, Suez Canal University, Ismailia 41522, Egypt; amr.elkelish@science.suez.edu.eg

* Correspondence: dilfuzajabborova@yahoo.com

Citation: Jabborova, D.; Wirth, S.; Halwani, M.; Ibrahim, M.F.M.; Azab, I.H.E.; El-Mogy, M.M.; Elkelish, A. Growth Response of Ginger (*Zingiber officinale*), Its Physiological Properties and Soil Enzyme Activities after Biochar Application under Greenhouse Conditions. *Horticulturae* **2021**, *7*, 250. https://doi.org/10.3390/horticulturae7080250

Academic Editor: Rui Manuel Almeida Machado

Received: 22 July 2021
Accepted: 12 August 2021
Published: 17 August 2021

Publisher's Note: MDPI stays neutral with regard to jurisdictional claims in published maps and institutional affiliations.

Abstract: This study aimed to investigate the effects of biochar (1%, 2%, and 3%) on seed germination, plant growth, root morphological characteristics, and physiological properties of ginger (*Zingiber officinale*) and soil enzymatic activities. Pot experiments under greenhouse conditions at 24 °C (day) and 16 °C (night) showed after six weeks that biochar additions of 2% and 3% significantly increased seed germination, plant height, leaf length, leaf number, as well as shoot and root dry weights compared to the control. Total root length significantly increased by 30%, 47%, and 74%, with increasing biochar contents (1%, 2%, and 3%) compared to the control. Root surface area, projected area, root diameter, and root volume reached a maximum at the 3% biochar treatment. The treatment with 2% biochar significantly increased fluorescein diacetate hydrolase and phenoloxidase activities by 33% and 59% compared to the control; so did the addition of 3% biochar, which significantly increased fluorescein diacetate hydrolases, phenoloxidase, and acid and alkaline phosphomonoesterase activity in soil compared to the control. Treatment with 3% biochar increased relative water content by 8%, chlorophyll content by 35%, and carotenoid content by 43% compared to the control. These results suggest that biochar can improve the performance of the rhizome of ginger and increase the activity of soil enzymes, thereby improving soil nutrient supply.

Keywords: Ginger (*Zingiber officinale*); biochar; plant growth; root morphological traits; chlorophyll content; soil enzymes

1. Introduction

Ginger (*Zingiber officinale* Rosc.) is an important herb and spice plant belonging to the Zingiberaceae family. The rhizome of ginger is significant for health and is considered effective against several ailments or disease-related manifestations such as headaches, nausea, vomiting, and motion sickness [1–3]. In addition, ginger is attributed with antitumorigenic and immunomodulatory effects. Furthermore, it serves as an antimicrobial, antiviral agent, is considered a potent analgesic and stimulant, and controls various diseases such as high cholesterol and blood pressure [4,5]. Ginger is used to treat a wide range of ailments such as stomach pain, diarrhea, nausea, asthma, and respiratory diseases [6]. The production of ginger as a nutrient-exhaustive crop requires an adequate supply of nutrients [7].

Mineral fertilizers and also biochars of various origins and qualities have been used to increase the yield of crops. Biochar contributes to decrease global warming, reduces atmospheric CO_2 concentrations, increases soil organic carbon (SOC), and improves soil nutrients [8,9]. Soil organic carbon increases after biochar application [10]. Accordingly, Scisłowska et al. [11] reported that biochar treatments improved the quality and productivity of soils. The biochar treatments significantly increased water holding capacity, cation exchange capacity, and specific surface area [12]. Biochar has also been reported to positively affect plant growth, development, and yield of various plants. Several reports have explicitly found that biochar increased seed germination, plant growth, and yield of various plants [13–19]. Thus, biochar treatment increased root biomass and shoot biomass of *Plantago lanceolate* compared to the control [20]. Rice straw biochar significantly increased plant height, bolls per plant, average boll weight, and cotton yield compared to the control treatment [21].

The addition of biochar can improve the availability of plant nutrients as well as plant physiological properties. The concentrations of Ca and Mg in corn leaf samples were significantly higher at a high biochar application rate than in the control [13]. Several studies have shown that biochar application increases plant photosynthesis, chlorophyll content, and transpiration rate [15,16,22]. The addition of biochar significantly increased the photosynthetic rate of okra (*Abelmoschus esculentus* L.) [18]. Regarding soil biological processes, it has been reported that the availability of nutrients and the activity of soil enzymes and microbial biomass in the soil are influenced by biochar. Accordingly, the availability of soil nutrients such as K, Ca, Mg, Na, and total C was improved by biochar [23,24]. In addition, biochar application significantly promoted N content in studies by Saxena et al. [25]. Moreover, biochar treatment has been reported to increase enzyme activities such as proteases, phosphohydrolases, and esterases [16,26,27]. Biochar application is accompanied by increased microbial biomass carbon content [28]. However, the effects of biochar or biochar-based fertilizers on ginger yield and soil nutrient availability have not been widely examined [29]. The objective of this study was to investigate the effect of biochar application on the growth, root morphological characteristics, physiological properties, and soil enzymatic activities of ginger (*Zingiber officinale*) grown under greenhouse conditions.

2. Materials and Methods

The biochar used in the study was produced at 450 °C from black cherry wood (Terra Anima® Onlineshop, Meissen, Germany) and had a particle size of less than 4 mm. The biochar and soil properties are summarised in Table 1 [19]. Ginger seeds (*Zingiber officinale*) were purchased from a local market in Berlin, Germany. The effect of biochar content on ginger growth was investigated in pot experiments in a greenhouse at ZALF, Müncheberg, Germany. All experiments were conducted in a randomized block design with three replicates. The experimental treatments included the control (soil without biochar) and soil with three levels of biochar (1%, 2%, and 3%). Plants were grown under greenhouse conditions at 24 °C during the day and 16 °C at night for 6 weeks. The seeds were cultivated in plastic pots (12 cm diameter, 18 cm depth) with 1.5 kg soil. Each pot was watered every 3 days. At harvest after 6 weeks, the germination rate, plant height, leaf length, leaf number, and leaf width, root and shoot fresh weight, and root and shoot dry weight were measured. The roots were carefully washed with water. The entire root system was then spread out and analysed with a scanning system (Expression 4990, Epson, Los Alamitos, CA) using a blue board as background. The digital images of the root system were analysed using Win RHIZO software (Régent Instruments, Quebec, QC, Canada). The total root length, root surface area, root volume, projected area, and root diameter were evaluated.

Table 1. Soil and biochar characteristics.

	Total (g kg^{-1})			Available (mg kg^{-1})				C/N	pH (H$_2$O)
	C	N	S	Ca	K	Mg	P		
Biochar	415.0	3.75	0.58	8893	1151	471	326	110.6	8.41
Soil	9.18	0.99	0.23	2103	1080	954	419	9.26	6.26

The relative water content (RWC) was measured according to the method of Barrs and Weatherley [30]. One hundred mg of fully expanded fresh leaf samples (FW) were placed in Petri dishes filled with double-distilled water for 4 h at room temperature immediately after sampling. The samples were then removed, blotted dry, and the threshold weight (TW) was recorded. The samples were then stored overnight in an oven at 70 °C, and the dry weight (DR) was recorded. The relative water content was calculated as:

$$RWC\ (\%) = [(FW - DW)/(TW - DW)] \times 100$$

Photosynthetic pigments were determined according to a modified method of Hiscox and Israelstam [31]. Freshly cut leaf samples of fifty mg pieces each, 2 to 3 mm in size, were placed in test tubes containing 5 mL DMSO. The test tubes were then incubated at 37 °C for 4 h in the dark. Incubation was continued until the tissue was completely colourless. The absorbance of the extract was measured at 470 nm, 645 nm, and 663 nm with a spectrophotometer against DMSO blank. The contents of chlorophyll a (Chl a), chlorophyll b (Chl b), total chlorophyll and carotenoids were determined using the following equation:

$$Total\ Chl\ (mg/g) = [20.2\ (A645) + 8.02\ (A663)] \times V/W$$

$$Carotenoids\ (mg/g) = [(1000 \times A470) - (3.27 \times Chl\ a + 104 \times Chl\ b)] \times V/W$$

where A = optical density; V = volume of DMSO (in ml); W = sample weight.

The acid and alkaline phosphatase activities were determined according to the method of Tabatabai and Bremner [32]. The hydrolytic activity of FDA was determined according to the method of Green et al. [33]. For this, a total of 0.5 mg of soil was mixed with 25 mL of sodium phosphate (0.06 M; pH 7.6). Then, 0.25 mL of a 4.9 mM FDA substrate solution was added to all test vials. All vials were mixed and incubated for 2 h in a water bath at 37 °C. The bottom suspension was then centrifuged at 8000 rpm for 5 min. The clear supernatant was measured at 490 nm against a reagent blank solution in a spectrophotometer (SpectraMax Plus 384).

Phenol oxidase (PO) activity was determined according to the method of Floch et al. [34]. Here, a modified universal buffer (MUB) stock solution was prepared according to Tabatabai [35], by dissolving 12.1 g tris(hydroxymethyl)aminomethane (THAM), 11.6 g maleic acid, 14.0 g citric acid, and 6.3 g boric acid in 488 mL 1 M sodium hydroxide (NaOH) and diluting the solution to 1 L with double-distilled water. Then 200 mL of the stock MUB solution was titrated to the desired pH with 0.1 M hydrochloric acid (HCl) or 0.1 M NaOH, and the volume was made up to 1 L with double-distilled water. An ABTS (2,2′- azinobis-(-3ethylbenzothiazoline-6-sulfononic acid) diammonium salt stock solution was prepared by dissolving 0.548 g ABTS in 10 mL double-distilled water for a final concentration of 0.1 M ABTS. PO activity was measured spectrophotometrically (SpectraMax Plus 384) with ABTS as substrate. The reaction mixture contained: 1.0 g soil, 10 mL MUB solution pH 4.0, and 200 μL of a 0.1 M ABTS solution. The final ABTS concentration in the incubation mixture was 2 mM. After incubation at 30 °C for 5 min, the mixture was centrifuged at 12,000 rpm for 2 min and the rate of oxidation of ABTS to ABTS+- released in the supernatant was measured at 420 nm.

The experimentally determined data were analysed with StatView software using ANOVA. The significance of the treatment effect was determined by the magnitude of the F value ($p < 0.05 < 0.001$).

3. Results

3.1. Plant Growth and Root Morphological Traits of Ginger

Biochar application generally increased seed germination compared to the control. Seed germination increased 8 days faster with the application of 2% and 3% biochar compared to the control (Figure 1). Both levels of biochar amendment (2% and 3%) significantly increased seed germination up to 95–98% on day 8 compared to the control. All three levels of biochar amendments increased seed germination to 100% on day 11 compared to the control.

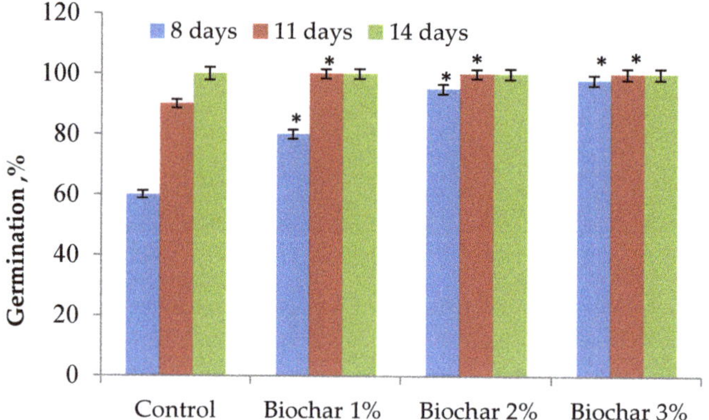

Figure 1. The impact of biochar concentrations on germination of ginger. Data are means of three replicates ($n = 3$), asterisks indicate significant differences from the control at $p < 0.05$. The 2% biochar treatment increased plant height by 53% compared to the control, and the 3% biochar treatment showed a 78% increase compared to the control treatment. The 2% and 3% biochar treatments also showed that leaf length increased by 36% and 57%, respectively, compared to the control. The leaf number increased significantly with increasing addition of biochar up to 85% compared to the control (Table 2). The leaf width also improved significantly with the addition of 2% and 3% biochar by 54% and 73% as compared to the control.

Table 2. The impact of biochar concentrations on plant height, leaf length, leaf number, and leaf width of ginger.

Treatments	Plant Height (cm)	Leaf Length (cm)	Leaf Number	Leaf Width (cm)
Control	18.0 ± 0.27	5.8 ± 0.01	3.3 ± 0.01	1.1 ± 0.01
Biochar 1%	22.3 ± 0.22	6.9 ± 0.03	4.0 ± 0.02	1.5 ± 0.01 *
Biochar 2%	27.5 ± 0.18 *	7.9 ± 0.01 *	4.8 ± 0.1 *	1.7 ± 0.01 *
Biochar 3%	32.1 ± 0.72 **	9.1 ± 0.03 *	6.1 ± 0.02 **	1.9 ± 0.01 **

Data are means of three replicates ($n = 3$), asterisks indicate significant differences from the control at $p < 0.05$, <0.01.

The shoot and root dry weights showed that the biochar treatments improved both the shoot and the root dry weight compared to the control (Figure 2). The 1% biochar treatment significantly increased shoot dry weight by 51% compared to the control. Treatment with 2% and 3% biochar significantly increased shoot dry weight by 67% and 79% compared to control. Root dry weight also increased sharply with increasing biochar, with a significant 67% increase at 2% biochar compared to the control (Figure 2). Root dry weight reached a maximum at 3% biochar treatment, with root dry weight significantly improved by 79% compared to the control.

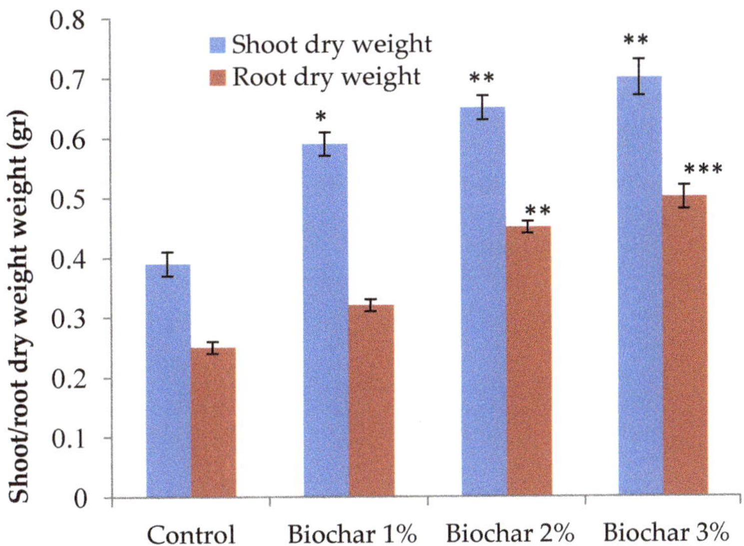

Figure 2. The impact of biochar concentrations on shoot and root dry weight of ginger. Data are means of three replicates ($n = 3$), asterisks indicate significant differences from the control at $p < 0.05$, <0.01, <0.001.

The data regarding the morphological characteristics of the roots showed that the total root length, root surface area, projected area, root diameter, and root volume increased with increasing addition (1%, 2%, and 3%) of biochar compared to the control (Table 3). Thus, total root length significantly increased by 30%, 47%, and 74% compared to the control, and root surface area significantly increased by up to 90% compared to the control (Table 3). The projected area reached a maximum at the 3% biochar treatment with an increase of 95% compared to the control. Root diameter improved greatly with the increasing amount of biochar by up to 97% compared to the control, as did root volume, which increased by up to 88% compared to the control (Table 3).

Table 3. The impact of biochar concentrations on root morphological traits of ginger.

Treatments	Total Root Length (cm)	Root Surface Area (cm^2)	Projected Area (cm^2)	Root Diameter (mm)	Root Volume (cm^3)
Control	114.1 ± 2.05	30.9 ± 1.12	23.0 ± 0.11	1.06 ± 0.01	1.17 ± 0.01
Biochar 1%	149.5 ± 2.83 *	46.7 ± 1.09 *	29.7 ± 0.13	1.20 ± 0.01	1.43 ± 0.01
Biochar 2%	167.8 ± 10.11 *	50.3 ± 2.00 **	38.8 ± 1.00 **	1.62 ± 0.02 *	1.68 ± 0.02 *
Biochar 3%	198.4 ± 12.00 **	58.6 ± 2.08 ***	44.8 ± 1.07 ***	2.09 ± 0.02 ***	1.99 ± 0.01 ***

Data are means of three replicates ($n = 3$), asterisks indicate significant differences from the control at $p < 0.05$, <0.01, <0.001.

3.2. Physiological Properties of Ginger

The data also showed that, depending on the concentration, the biochar treatments also improved the relative water content compared to the control. With increasing addition of biochar, the water content increased significantly by up to 8% compared to the control after 3% biochar amendment (Figure 3).

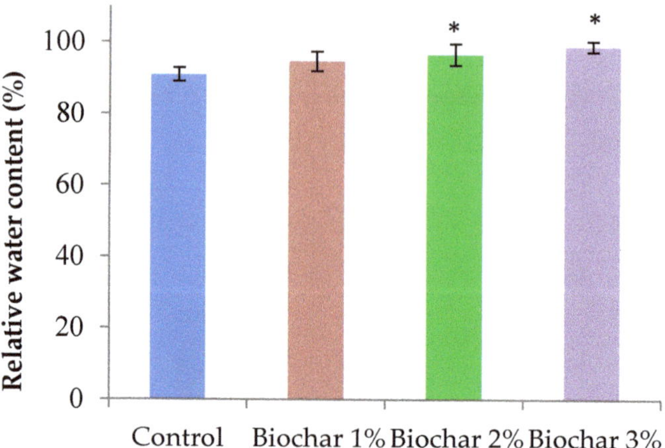

Figure 3. The impact of biochar concentrations on the relative water content of ginger. Data are means of three replicates ($n = 3$), asterisks indicate significant differences from the control at $p < 0.05$.

The data showed that the 2% biochar treatment significantly increased chlorophyll content by 23% compared to the control. The 3% biochar treatment showed a further increasing positive effect by 35% compared to the control (Figure 4).

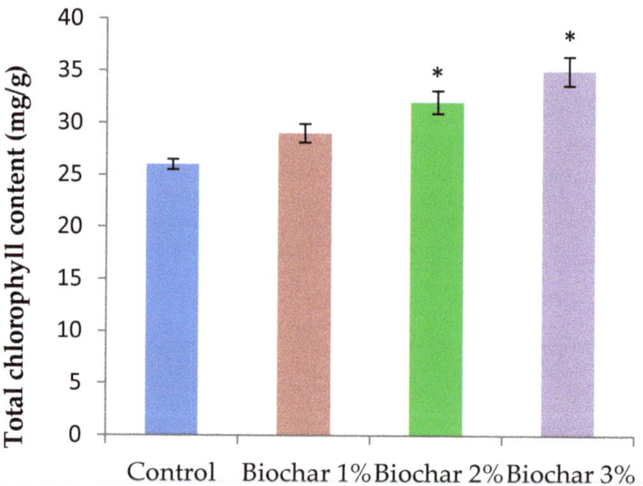

Figure 4. The impact of biochar concentrations on chlorophyll content of ginger. Data are means of three replicates ($n = 3$), asterisks indicate significant differences from the control at $p < 0.05$.

In addition, the results on carotenoid contents showed that the biochar additions of 2% and 3% significantly increased the carotenoid content by 30% and 43%, respectively, compared to the control (Figure 5).

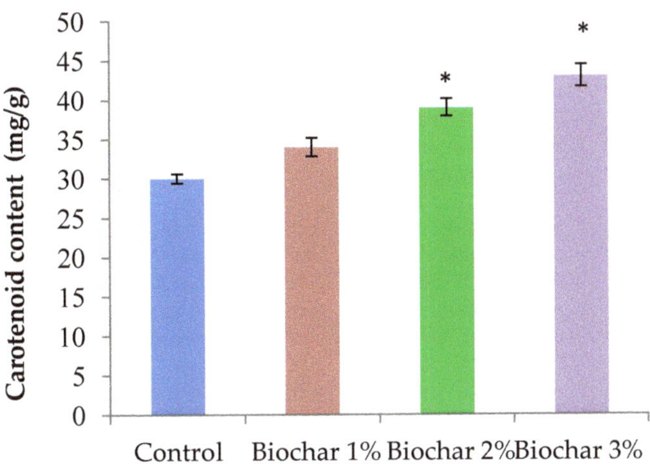

Figure 5. The impact of biochar concentrations on carotenoid content of ginger. Data are means of three replicates ($n = 3$), asterisks indicate significant differences from the control at $p < 0.05$.

3.3. Soil Enzyme Activities

The addition of biochar increased the activities of fluorescein diacetate hydrolases, phenol oxidases, and acidic and alkaline phosphomonoestrases in soil. The data in Table 4 show that 1% biochar treatment significantly increased phenol oxidase activity by 38% compared to the control, and the 2% addition significantly increased fluorescein diacetate hydrolase activity and phenol oxidase activity by 33% and 59% compared to the control. Soil acidic and alkaline phosphomonoestrase activity reached a maximum at the 3% biochar treatment compared to the control. The 3% addition of biochar increased the fluorescein diacetate hydrolase activity as well as the phenol oxidase activity by 55% and 77%, respectively, compared to the control.

Table 4. The impact of biochar concentrations on soil enzymes.

Treatments	Acid Phosphomono-Esterase ($\mu g\ g^{-1}\ h^{-1}$)	Alkaline Phosphomono-Esterase ($\mu g\ g^{-1}\ h^{-1}$)	FDA (Fluorescein Diacetate) Activity ($\mu g\ g^{-1}\ h^{-1}$)	Phenol Oxidase Activity (U.g^{-1} DW)
Control	980.1 ± 36.6	656.7 ± 10.5	40.6 ± 1.19	22.1 ± 0.11
Biochar 1%	1099.8 ± 38.4	677.1 ± 12.6	45.7 ± 1.30	30.5 ± 0.17 *
Biochar 2%	1176.9 ± 43.0	698.2 ± 15.3	54.2 ± 1.12 *	35.2 ± 1.10 *
Biochar 3%	1238.4 ± 44.1 *	753.6 ± 20.1 *	62.8 ± 1.28 *	39.2 ± 1.02 **

Data are means of three replicates ($n = 3$), asterisks indicate significant differences from the control at $p < 0.05$, <0.01.

4. Discussion

In the present study, the control treatment reduced seed germination, plant height, leaf number, leaf length, leaf width of ginger. Accordingly, several studies have reported that ginger growth and yield significantly decreased without mineral fertilizer [36–40], proving that ginger is a plant with specific nutrient requirements.

In the present study, we used black cherry biochar at different concentrations to enhance seed germination and growth of ginger. Together, the concentrations of 2% and 3% biochar promoted maximum seed germination, which was significantly higher compared to the control. Accordingly, Kanwal et al. [41] reported that the addition of 1% and 2% biochar increased the germination of wheat seeds. Moreover, this result is in agreement with the report of Bu et al. [42], who observed a significant increase in germination rate, the dry matter of shoots and roots of *Robinia pseudoacacia* L. seeds by rice husk biochar and wood chips biochar. Similar results confirming improved germination of castor seeds by the addition of 1% and 5% castor stem biochar were reported by Hilioti et al. [43]. The positive effect of biochar addition on seed germination, root length, shoot length, root biomass, shoot biomass, and yield has also been reported by Saxena et al. [25]. Jabborova et al. [44] reported that biochar application significantly increased shoot length, leaf lengths, leaf number, leaf width, root dry weight, shoot dry weight, and total root length. This result is in agreement with the report of Carter et al. [14], who observed that application of biochar from rice husks increased final biomass and root biomass of lettuce (*Lactuca sativa*) and cabbage (*Brassica chinensis*) compared to treatment without biochar. In terms of plant height, leaf length, leaf number, width shoot dry weight, and root dry weight, it was shown that the 2% and 3% biochar treatments significantly increased these plant growth parameters compared to the control. Thus, several other researchers reported that biochar increased plant growth, root dry matter, shoot dry matter and yield of numerous other crops [13,15,17,20,45].

Root morphological parameters such as total root length, root surface area, projected area, root diameter, and root volume were improved by biochar. In the present study, root morphological characteristics such as total root length, root surface area, projected area, root diameter, and root volume were observed to be improved by biochar additions of 2% and 3% compared to the control. A number of other studies have also reported that biochar application improved plant root growth and development [17,27,46]. A significant increase in root length, root surface area, and root volume at an application rate of 1% rice husk biochar and woodchip biochar has also been reported by Bu et al. [42]. The results of this study showed that the 3% biochar treatment significantly increased root diameter and projected area compared to all treatments. In the present study, ginger physiological properties such as chlorophyll content, carotenoid content, and relative water content were indicated to be improved by biochar additions of 2% and 3% compared to the control without biochar. Similar results were reported by other authors [12,15,16], where the biochar application increases plant photosynthesis, chlorophyll content, and transpiration rate. The application of biochar significantly increased the content of chlorophyll a, chlorophyll b, total chlorophyll, carotenoid contents, and relative water of spinach (*Spinacia oleracea* L.) [44]. Chrysargyris et al. [47] reported that trees biochar at 7.5% and 15% significantly increased total chlorophylls content of lettuce. Our study further showed that biochar treatment positively affected plant physiological parameters, such as chlorophyll content and carotenoid content, which reached a maximum of 3% biochar treatment compared to the control. Similarly, Sarma et al. [18] reported a pronounced positive effect of biochar addition on the photosynthetic rate of okra. Other researchers found that biochar application increased chlorophyll content, transpiration rate, total flavonoids, sugars, and glucose in various plants in addition to photosynthesis [15,22,48]. Corresponding results confirming a significant increase of 27.1% in photosynthetic rate and 16.1% in chlorophyll concentration by biochar addition were reported by He et al. [49]. The activity of a number of soil enzymes was promoted by biochar application. In the present study, fluorescein diacetate hydrolase, phenoloxidase, and soil acidic and alkaline phosphomonoesterase activities reached a maximum with 3% biochar treatment compared to the control and the other biochar treatments. Thus, other authors also reported that biochar applications increased soil enzymes such as proteases, phosphohydrolase, lipases, and esterases [16,26,27]. This result also confirms studies by Bailey et al. [50] and Ma et al. [19], both of which observed increased soil enzyme activity upon biochar application, suggesting increased soil microbiological activity. Similar results were confirmed after biochar application by Wang et al. [51]. A significant increase in urease activity by 40%, invertase activity by 9%, and phosphatase activity by 46% with biochar application has been reported by Oladele [52]. Thus, increased activity of soil enzymes due to the addition of biochar also contributed to better availability of nutrients in the soil [53].

5. Conclusions

This study found that biochar application improved seed germination, plant height, leaf length, leaf number, leaf width, as well as shoot and root weight of ginger. The most effective treatment of 3% biochar application also significantly increased root dry weight, shoot dry weight, and root morphological characteristics. The 3% biochar treatment also had a pronounced, positive impact on soil enzyme activity, relative water content, chlorophyll content and carotenoid content as compared to all other treatments. Thus, the present study indicates that biochar application is a promising option to improve and stabilize soil fertility, yield, and quality criteria in ginger cultivation.

Author Contributions: Conceptualization, D.J. and S.W.; methodology, D.J. and M.H.; software, D.J., I.H.E.A., M.M.E.-M. and A.E.; validation, D.J., I.H.E.A. and M.M.E.-M.; formal analysis, D.J., A.E. and M.F.M.I.; investigation, D.J., I.H.E.A., M.M.E.-M., A.E. and M.F.M.I.; writing—original draft preparation, D.J.; writing—review and editing, D.J., S.W., I.H.E.A., M.M.E.-M., A.E. and M.F.M.I. All authors have read and agreed to the published version of the manuscript.

Funding: Work stay in Germany was funded by the German Academic Exchange Service DAAD Re-invitation Programme for Former Scholarship Holders, 2019 (57440916).

Data Availability Statement: Not applicable.

Acknowledgments: The authors extend their appreciation to Deanship of Scientific Research at Taif University for funding this work through Researchers Supporting Project number TURSP-2020/27, Taif University, Taif, Saudi Arabia. We thank colleagues at the Leibniz Centre for Agricultural Landscape Research (ZALF), Müncheberg, Germany, for providing support in the laboratory and greenhouse facilities, namely, the Experimental Field Station and the Central Laboratory.

Conflicts of Interest: The authors declare no conflict of interest.

References

1. Dedov, V.N.; Tran, V.H.; Duke, C.C.; Connor, M.; Christie, M.J.; Mandadi, S.; Roufogalis, B.D. Gingerols: A novel class of vanilloid receptor (VR1) agonists. *Br. J. Pharmacol.* **2002**, *137*, 793–798. [CrossRef] [PubMed]
2. Borrelli, F.; Capasso, R.; Pinto, A.; Izzo, A.A. Inhibitory effect of ginger (*Zingiber officinale*) on rat ileal motility in vitro. *Life Sci.* **2004**, *74*, 2889–2896. [CrossRef]
3. Newerli-Guz, J.; Pych, A. Antixodiant properties of ginger (*Zingiber officinale Roscoe*). *Zesz. Nauk. AMG* **2012**, *73*, 28–33.
4. Vimala, S.; Norhanom, A.W.; Yadav, M. Anti-tumor promoter activity in Malaysian ginger rhizome used in traditional medicine. *Br. J. Cancer* **1999**, *80*, 110–116. [CrossRef]
5. Akoachere, J.F.; Ndip, R.N.; Chenwi, E.B.; Ndip, L.M.; Njock, T.E.; Anong, D.N. Antibacterial effect of *Zingiber officinale* and Garcinia kola on respiratory tract pathogens. *East Afr. Med. J.* **2002**, *79*, 588–592. [CrossRef] [PubMed]
6. Grzanna, R.; Lindmark, L.; Frondoza, C. Ginger-A herbal medicinal product with broad anti-inflammatory actions. *J. Med. Food* **2005**, *8*, 125–132. [CrossRef]
7. Jabborova, D.; Sayyed, R.Z.; Azimov, A.; Jabbarov, Z.; Matchanov, A.; Enakiev, Y.; Baazeem, A.; Sabagh, A.E.; Danish, S.; Datta, R. Impact of mineral fertilizers on mineral nutrients in the ginger rhizome and on soil enzymes activities and soil properties. *Saudi J. Biol. Sci.* **2021**, in press. [CrossRef]
8. Lehmann, J.; Skjemstad, J.O.; Sohi, S.; Carter, J.; Barson, M.; Falloon, P.; Coleman, K.; Woodbury, P.; Krull, E. Australian climate-carbon cycle feedback reduced by soil black carbon. *Nat. Geosci.* **2008**, *1*, 832–835. [CrossRef]
9. Stanton, C.Y.; Mach, K.J.; Turner, P.A.; Lalonde, S.J.; Sanchez, D.L.; Field, C.B. Managing cropland and rangeland for climate mitigation: An expert elicitation on soil carbon in California. *Clim. Chang.* **2018**, *147*, 633–646. [CrossRef]
10. Suddick, E.C.; Six, J. An estimation of annual nitrous oxide emissions and soil quality following the amendment of high temperature walnut shell biochar and compost to a small scale vegetable crop rotation. *Sci. Total. Environ.* **2013**, *465*, 298–307. [CrossRef]
11. Scisłowska, M.; Włodarczyk, R.; Kobyłecki, R.; Bis, Z. Biochar to improve the quality and productivity of soils. *J. Ecol. Eng.* **2015**, *16*, 31–35. [CrossRef]
12. Laird, D.A.; Fleming, P.; Davis, D.D.; Horton, R.; Wang, B.; Karlen, D.L. Impact of biochar amendments on the quality of a typical Midwestern agricultural soil. *Geoderma* **2010**, *158*, 443–449. [CrossRef]
13. Major, J.; Rondon, M.; Molina, D.; Riha, S.J.; Lehmann, J. Maize yield and nutrition during 4 years after biochar application to a Colombian savanna oxisol. *Plant Soil* **2010**, *333*, 117–128. [CrossRef]
14. Carter, S.; Shackley, S.; Sohi, S.; Suy, T.B.; Haefele, S. The Impact of Biochar Application on Soil Properties and Plant Growth of Pot Grown Lettuce (*Lactuca sativa*) and Cabbage (*Brassica chinensis*). *Agronomy* **2013**, *3*, 404–418. [CrossRef]

15. Agegnehu, G.; Bass, A.M.; Nelson, P.N.; Muirhead, B.; Wright, G.; Bird, M.I. Biochar and biochar-compost as soil amendments: Effects on peanut yield, soil properties and greenhouse gas emissions in tropical North Queensland, Australia. *Agric. Ecosyst. Environ.* **2015**, *213*, 72–85. [CrossRef]

16. Trupiano, D.; Cocozza, C.; Baronti, C.; Amendola, C.; Vaccari, F.P.; Lustrato, G.; Lonardo, S.D.; Fantasma, F.; Tognetti, R.; Scippa, G.S. The Effects of Biochar and Its Combination with Compost on Lettuce (*Lactuca sativa* L.) Growth, Soil Properties, and Soil Microbial Activity and Abundance. *Int. J. Agron.* **2017**, *2017*, 3158207. [CrossRef]

17. Głodowska, M.; Schwinghamer, T.; Husk, B.; Smith, D. Biochar based inoculants improve soybean growth and nodulation. *Agric. Sci.* **2017**, *8*, 1048–1064. [CrossRef]

18. Sarma, B.; Borkotoki, B.; Narzari, R.; Kataki, R.; Gogoi, N. Organic amendments: Effect on carbon mineralization and crop productivity in acidic soil. *J. Clean. Prod.* **2017**, *152*, 157–166. [CrossRef]

19. Ma, H.; Egamberdieva, D.; Wirth, S.; Bellingrath-Kimura, S.D. Effect of biochar and irrigation on soybean-Rhizobium symbiotic performance and soil enzymatic activity in field rhizosphere. *Agronomy* **2019**, *9*, 626. [CrossRef]

20. Zhaoxiang, W.; Huihu, L.; Qiaoli, L.; Changyan, Y.; Yu, F. Application of bio-organic fertilizer, not biochar, in degraded red soil improves soil nutrients and plant growth. *Rhizosphere* **2020**, *16*, 100264. [CrossRef]

21. Qayyum, M.F.; Haider, G.; Raza, M.A.; Abdel, K.S.H.; Mohamed, A.K.S.N.; Rizwan, M.; El-Sheikh, M.A.; Alyemeni, M.N.; Ali, S. Straw-based biochar mediated potassium availability and increased growth and yield of cotton (*Gossypium hirsutum* L.). *J. Saudi Chem. Soc.* **2020**, *24*, 963–973. [CrossRef]

22. Speratti, A.B.; Johnson, M.S.; Sousa, H.M.; Dalmagro, H.; Couto, E.G. Biochars from local agricultural waste residues contribute to soil quality and plant growth in a Cerrado region (Brazil) Arenosol. *GCB Bioenergy* **2018**, *10*, 272–286. [CrossRef]

23. Wang, Y.; Yin, R.; Liu, R. Characterization of biochar from fast pyrolysis and its effect on chemical properties of the tea garden soil. *J. Anal. Appl. Pyrol.* **2014**, *110*, 375–381. [CrossRef]

24. Gaskin, J.W.; Speir, R.A.; Harris, K.; Das, K.C.; Lee, R.D.; Morris, L.A.; Fisher, D.S. Effect of peanut hull and pine chip biochar on soil nutrients, corn nutrient status, and yield. *Agron. J.* **2010**, *102*, 623–633. [CrossRef]

25. Saxena, J.; Rana, G.; Pandey, M.J. Impact of addition of biochar along with *Bacillus* sp. on growth and yield of French beans. *Sci. Hortic.* **2013**, *162*, 351–356. [CrossRef]

26. Anderson, C.R.; Condron, I.M.; Clough, T.J.; Fiers, M.; Stewart, A.; Hill, R.A.; Sherlock, R.R. Biochar induced soil microbial community change: Implications for biogeochemical cycling of carbon, nitrogen and phosphorus. *Pedobiologia* **2011**, *54*, 309–320. [CrossRef]

27. Jabborova, D.; Wirth, S.; Kannepalli, A.; Narimanov, A.; Desouky, S.; Davranov, K.; Sayyed, R.Z.; El Enshasy, H.; Malek, R.A.; Syed, A.; et al. Co-Inoculation of Rhizobacteria and Biochar Application Improves Growth and Nutrientsin Soybean and Enriches Soil Nutrients and Enzymes. *Agronomy* **2020**, *10*, 1142. [CrossRef]

28. Zhou, H.M.; Zhang, D.X.; Wang, P.; Liu, X.Y.; Cheng, K.; Li, L.; Zheng, J.; Zhang, X.; Zheng, J.; Crowley, D.; et al. Changes in microbial biomass and the metabolic quotient with biochar addition to agricultural soils: A meta-analysis. *Agric. Ecosyst. Environ.* **2017**, *239*, 80–89. [CrossRef]

29. Farrar, M.B.; Wallace, H.M.; Xu, C.Y.; Nguyen, T.T.; Tavakkoli, E.; Joseph, S.; Bai, S.H. Short-term effects of organo-mineral enriched biochar fertiliser on ginger yield and nutrient cycling. *J. Soils Sediments* **2019**, *19*, 668–682. [CrossRef]

30. Barrs, H.D.; Weatherley, P.E. A reexamination of the relative turgidity technique for estimating water deficits in leaves. *Aust. J. Biol. Sci.* **1962**, *15*, 413–428. [CrossRef]

31. Hiscox, J.; Israelstam, G. A method for the extraction of chlorophyll from leaf tissue without maceration. *Can. J. Bot.* **1979**, *57*, 1332–1334. [CrossRef]

32. Tabatabai, M.A.; Bremner, J.M. Use of p-nitrophenol phosphate for the assay of soil phosphatase activity. *Soil Biol. Biochem.* **1969**, *1*, 301–307. [CrossRef]

33. Green, V.S.; Stott, D.E.; Diack, M. Assay for Fluorescein Diacetate Hydrolytic Activity: Optimization for Soil Samples. *Soil Biol. Biochem.* **2006**, *38*, 693–701. [CrossRef]

34. Floch, C.; Gutiérrez, E.A.; Criquet, S. ABTS assay of phenol oxidase activity in soil. *J. Microbiol. Methods* **2007**, *71*, 319–324. [CrossRef]

35. Tabatabai, M.A. Soil enzymes. In *Methods of Soil Analysis: Microbiological and Biochemical Properties*; Weaver, R.W., Angle, J.S., Bottomley, P.S., Eds.; Soil Science Society of America: Madison, WI, USA, 1994; pp. 775–833.

36. Jabborova, D.; Choudhary, R.; Karunakaran, R.; Ercisli, S.; Ahlawat, J.; Sulaymanov, K.; Azimov, A.; Jabbarov, Z. The Chemical Element Composition of Turmeric Grown in Soil–Climate Conditions of Tashkent Region, Uzbekistan. *Plants* **2021**, *10*, 1426. [CrossRef] [PubMed]

37. Kun, X.; Feng, X. Effect of nitrogen on the growth and yield of ginger. *China Veg.* **1999**, *6*, 12–14.

38. Haque, M.M.; Rahman, A.K.; Ahmed, M.; Masud, M.M.; Sarker, M.M. Effect of nitrogen and potassium on the yield and quality of ginger in hill slope. *J. Soil Nat.* **2007**, *1*, 36–39.

39. Singh, M.; Khan, M.M.; Naeem, M. Effect of nitrogen on growth, nutrient assimilation, essential oil content, yield and quality attributes in *Zingiber officinale* Rosc. *J. Saudi Soc. Agric. Sci.* **2016**, *15*, 171–178. [CrossRef]

40. Jabborova, D.; Enakiev, Y.; Sulaymanov, K.; Kadirova, D.; Ali, A.; Annapurna, K. Plant growth promoting bacteria *Bacillus subtilis* promote growth and physiological parameters of *Zingiber officinale* Roscoe. *Plant Sci. Today* **2021**, *8*, 66–71. [CrossRef]

41. Kanwal, S.; Ilyas, N.; Shabir, S.; Saeed, M.; Gul, R.; Zahoor, M.; Batool, N.; Mazhar, R. Application of biochar in mitigation of negative effects of salinity stress in wheat (*Triticum aestivum* L.). *J. Plant Nutr.* **2018**, *41*, 526–538. [CrossRef]
42. Bu, X.; Xue, J.; Wu, Y.; Ma, W. Effect of Biochar on Seed Germination and Seedling Growth of *Robinia pseudoacacia* L. In Karst Calcareous Soils. *Commun. Soil Sci. Plant Anal.* **2020**, *51*, 352–363. [CrossRef]
43. Hilioti, Z.; Michailof, C.M.; Valasiadis, D.; Iliopoulou, E.F.; Koidou, V.; Lappas, A.A. Characterization of castor plant-derived biochars and their effects as soil amendments on seedlings. *Biomass Bioenergy* **2017**, *105*, 96–106. [CrossRef]
44. Jabborova, D.; Annapurna, K.; Paul, S.; Kumar, S.; Ibrahim, M.; Elkelish, A.A. Beneficial features of biochar and AMF for improving spinach plant growth, root morphological traits, physiological properties and soil enzymatic activities. *J. Fungi.* **2021**, *7*, 571. [CrossRef]
45. Agboola, K.; Moses, S. Effect of biochar and cowdung on nodulation, growth and yield of soybean (*Glycine max* L. Merrill). *Int. J. Agric. Biosci.* **2015**, *4*, 154–160.
46. Uzoma, K.; Inoue, M.; Andry, H.; Fujimaki, H.; Zahoor, A.; Nishihara, E. Effect of cow manure biochar on maize productivity under sandy soil condition. *Soil Use Manag.* **2011**, *27*, 205–212. [CrossRef]
47. Chrysargyris, A.; Prasad, M.; Kavanagh, A.; Tzortzakis, N. Biochar type, ratio, and nutrient levels in growing media affects seedling production and plant performance. *Agronomy* **2020**, *10*, 1421. [CrossRef]
48. Petruccelli, R.; Bonetti, A.; Traversi, M.L.; Faraloni, C.; Valagussa, M.; Pozz, A. Influence of biochar application on nutritional quality of tomato (*Lycopersicon esculentum*). *Crop. Pasture Sci.* **2015**, *66*, 747–755. [CrossRef]
49. He, Y.; Yao, Y.; Ji, Y.; Deng, J.; Zhou, G.; Liu, R.; Shao, J.; Zhou, L.; Li, N.; Zhou, X. Biochar amendment boosts photosynthesis and biomass in C3 but not C4 plants: A global synthesis. *GCB Bioenergy* **2020**, *12*, 605–617. [CrossRef]
50. Bailey, V.L.; Fansler, S.J.; Smith, J.L.; Bolton, H., Jr. Reconciling apparent variability in effects of biochar amendment on soil enzyme activities by assay optimization. *Soil Biol. Biochem.* **2011**, *43*, 296–301. [CrossRef]
51. Wang, L.; Xue, C.; Nie, X.; Liu, Y.; Chen, F. Effects of biochar application on soil potassium dynamics and crop uptake. *J. Plant Nutr. Soil Sci.* **2018**, *181*, 635–643. [CrossRef]
52. Oladele, S.O. Effect of biochar amendment on soil enzymatic activities, carboxylate secretions and upland rice performance in a sandy clay loam Alfisol of Southwest Nigeria. *Sci. Afr.* **2019**, *4*, e00107. [CrossRef]
53. Ouyang, L.; Tang, Q.; Yu, L.; Zhang, R. Effects of amendment of different biochars on soil enzyme activities related to carbon mineralisation. *Soil Res.* **2014**, *52*, 706–716. [CrossRef]

MDPI

St. Alban-Anlage 66

4052 Basel

Switzerland

Tel. +41 61 683 77 34

Fax +41 61 302 89 18

www.mdpi.com

Horticulturae Editorial Office

E-mail: horticulturae@mdpi.com

www.mdpi.com/journal/horticulturae

www.ingramcontent.com/pod-product-compliance
Lightning Source LLC
LaVergne TN
LVHW070620170726
843515LV00004B/133